REAL Science Odyssey

Biology **LEVEL 2**

Workbook

Blair H. Lee, MS

www.pandiapress.com

Biology LEVEL 2 Workbook

Table of Contents

Table of Contents

Chapter 1: All Living Things

Explore

What's Out There? Plot Study

Chapter 1: Lab

Plot studies help scientists monitor fluctuations of species in a specific area.

In 1980, two biologists, Robin Foster and Stephen Hubbell, began a **plot study** in the rain forest on Barro Colorado Island, located in Gatun Lake in Panama. One purpose of this experiment has been to monitor changes to the numbers and **species** (types) of trees in the area. The size of the plot they are studying is 50 hectares, or about 124 acres.

To begin their plot study they chose an area that represented the overall tree population on the island. After choosing the location, they measured the perimeter of the plot. Next, the scientists carefully walked through the area and plotted (mapped) the different tree species. They counted the numbers of each species. They found about 300 different species of trees in the 300,000 they counted! Since then, scientists regularly revisit the plot to monitor any changes. There are now over 18 of these types of plots and plot studies worldwide, monitoring about 6,000 tree species. The plots are called "earth observatories." They are in Asia, Africa, and Latin America.

Plot study experiments are one method of determining what's out there. Researchers map an area or plot. They use the data collected from the plot to estimate what is in a larger area. These 18 plots have taught scientists a lot about what's out there and what's happening in rain forests worldwide.

Today you are going to conduct your own plot study. You are not going to count 300,000 trees or walk over 124 acres, but you are going to do what Drs. Foster and Hubbell did. You are going to go to an area and find a small plot within that area to conduct a study. You will mark off the boundaries of that area and study what is living in it. Be sure to check in trees and under rocks. While you are at it, check to see if anything is growing on that rock. Check the plants in your plot for insects; maybe you will find caterpillar eggs or a cocoon. Take the time to look for both the small and the tall in the area you choose to study.

Chapter 1: Lab *(continued)*

Materials

- Tape measure
- Graph paper
- Notebook paper
- Data tables
- Clipboard
- Calculator

- Outdoor area to conduct plot study
- Field guide(s): be prepared to identify insects, plants, birds, reptiles, mammals, and if you choose a water area—fish

- 4 markers for the corners of your plot

Procedure

1. Write your hypothesis for this experiment on the Lab Report before performing it. For this lab, and a few other labs in the course, you will be writing a formal lab report. The following explains how to write a lab report:

> The scientific method is based on experimentation, observation, and deductive reasoning. Lab reports about the experiments use the scientific method. Before you write anything down about an experiment, you need to understand what you are testing. Ask yourself, "What question is this experiment asking?" When you can answer that, you are ready to make your **Hypothesis** (an educated guess) about the outcome of the experiment. The next section is the **Procedure**. In this section you will briefly rewrite the steps used to conduct the experiment. The **Observations** section is where you keep track of what happened during the experiment. The **Results and Calculations** section is where your data, calculations, and tables go. In the **Conclusions** section, take the information from the observations and results and calculations sections and use deductive reasoning to determine an answer to the question the experiment asks. If there are any weaknesses to the experiment, this is the place to state them.

2. Decide on the location of the plot. Are you going to map a forested area, desert area, your lawn, a slow-moving creek, or a playground? When you decide on an area, try to choose a plot that is a good representative of the area as a whole.

3. Use your tape measure or meter stick and measure a 2 meter by 2 meter (2m x 2m) square plot on the ground. Mark the boundaries of the plot with a marker at each corner.

4. In the center of a piece of graph paper, draw an outline of a large block that is 20 squares by 20 squares. This will be used to draw a representation of your plot. Your drawing of the plot will be *scaled*. Each square on the graph paper accounts for 10 centimeters (⅒ of a meter) of the plot. Therefore, the scale

Chapter 1: Lab *(continued)*

when drawing the plot is 10:1. This means that every 10 centimeters of the plot will be drawn into one square of graph paper. If your plot size varies, the size of your rectangle will be different. For example, if your plot is 1m x 2m, then draw a block that is 10 squares by 20 squares on the graph paper.

5. Draw the plot on your graph paper. Try to include a sketch of everything. Do your best to count the total number of each species of animal and plant. If there are rocks or other non-living objects that can be lifted, lift them very carefully, and on a separate sheet make notes about and drawings of what is underneath. When you are done, carefully replace the objects, so all the little animals still have a home.

6. Using your field guide, do your best to identify the plants and animals in the plot. If you can find the name, write it down.

7. You can fill in your data tables in the field or after you get home.

8. Note that some species of plants and animals might look different when they are really the same organism, depending on:

   - Their life stage. Caterpillars become butterflies and tadpoles become frogs in a process called **metamorphosis**. Animals that go through metamorphosis look different as adults from when they are babies.

   - Plants can look different depending on their life stage. Males and females of the same species of animal can look very different from one another. This is called **sexual dimorphism**. Take, for example, mallard ducks. The female is a drab brown all over and the male has a colorful green head with a white ring around its neck.

9. In addition to plants and animals, look for fungi (fungi is the plural form of fungus).

   - Mushrooms are fungi.

   - The orange/yellow/red growths on rocks are living fungi or lichens (fungi and algae).

When you have finished drawing the plot, on a sheet of notebook paper:

10. Describe the plot, in words.

11. Describe the entire area the plot represents. Some questions you might address:

   - Did you observe any wildlife nearby that was not included in your plot?

   - Did you include the one and only one tree in the area in your plot? In other words, if you did not choose a representative sample of the area for your plot, this is the place to note that.

   - Did you expect to find a species of wildlife that you didn't see?

12. Write your report using the following Lab Report forms. Refer to Lab Calculations as needed for assistance in writing your report.

What's Out There? Plot Study

Chapter 1: Lab Report

Name_____ Date_____

Title/Location_____

Hypothesis

Procedure

Observations

Results and Calculations

Conclusions

Pandia PRESS

Chapter 1: Lab Report–Data Tables

Table 1 (refer to Lab Calculations to complete this table)

Animals (list each species)	# of species in my plot	# of my plots that fit into 100m²	Estimate # of species for 100m²

Total animal species =

Notes:

Chapter 1: Lab Report–Data Tables *continued*

Table 2 (refer to Lab Calculations to complete this table)

Plants (list each species)	# of species in my plot	# of my plots that fit into 100m²	Estimate # of species for 100m²

Total plant species =

Table 3 (refer to Lab Calculations to complete this table)

	# Species found in your plot	Estimate # missed	Estimate # of species in 100m²
Animals			
Plants			
Fungi			

Pandia PRESS

Chapter 1: Lab *continued*

Lab Calculations

Use the following lab calculations to assist you in completing Lab Report–Data Tables. (It looks like there are a lot of steps for the math part. That is because I spend a lot of time explaining it.)

You measured and described a small part of a larger area. Using math, you are going to estimate how many organisms are in a bigger area and how many species you might have missed.

Tables 1 and 2
How many of each type of organism is in a 100-square-meter area, 100m²?

    A. Calculate the area of your plot.

        Area = length x width

        2m x 2m = _____m²

    B. Calculate how many of these plots would fit into 100m².

        100m² ÷ the answer from **A** = number of plots that would fit in 100m².

        100m² ÷ _____ = _____

This number is used to estimate how many of each type of organism is in the larger area.

Example: If you observed 5 beetles in your 2m x 2m plot, you would multiply the answer from **B** by 5, the number of beetles you saw. That is your estimate of the total number of beetles in 100m².

Animals (list each species)	# of species in my plot	# of my plots that fit into 100m²	Estimate # of species for 100m²
Beetle	5	25	125

Now it's your turn. Fill in Tables 1 and 2 with your calculated estimates for the numbers of plant and animal species in a 100m² area.

Table 3
Do you think every plant and animal in a 100m² area was included in the 2m x 2m plot? Most likely not; some error is introduced into a study like this because we are looking at a small area. You are going to calculate an estimate of error, also known as a fudge factor, to correct for this.

Chapter 1: Lab *continued*

Let's assume the 2m x 2m plot you chose did not include 10 percent of the plant and animal species found in the 100m² area. Because percent means "out of a hundred," this means ten out of every hundred plant and animal species in an area of 100m² were missed.

If 10 percent is converted to a decimal, you can use the decimal to find an estimate for the number of species missed.

$$10\% \div 100 = .1$$

Example: For every ten species of organisms included in your plot, how many were missing from the larger area? 10 x .1 = 1. There was one species in 100m² that was not in the 2m x 2m plot.

So let's say you found 17 animal species in your plot. To use this estimate to find the total number of animal species you would expect to find in the larger area:

17 x .1 = 1.7 (There cannot be .7 of a species, so you need to round to the nearest whole number.) 1.7 is rounded up to 2.

> A note about rounding numbers: If the decimal is .1 to .4, round down to the nearest whole number. If the decimal is .5 to .9, round up to the nearest whole number. For example: 1.1, 1.2, 1.3, and 1.4 are all rounded DOWN to 1. While 1.5, 1.6, 1.7, 1.8, and 1.9 are all rounded UP to 2.

You would estimate that two species of animals would be found in the larger area, 100m², that were not seen in the 2m x 2m plot.

How many animal species would you expect to find in 100m²?

Number in plot + number missed = total number.

17 + 2 = 19, this is your estimate for the total number of animal species in 100m², based on the results of your 2m x 2m plot study.

	# Species found in your plot	Estimate # missed	Estimate # of species in 100m²
Animals	17	2	19

Now it's your turn. Count all the different species of plants, animals, and fungi (if you found any) from your data tables. (If you saw 10 of the same species of spider, that would be 1, not 10.) Complete Table 3 by estimating the number of species you would find in a larger area.

Explore

Your Microscope: Parts
Chapter 1: Microscope Lab 1

The instructions for the microscope labs are long in the first few chapters. They are written for students who have never used a microscope. Even for those of you who have used a microscope before, there is a lot of information in these beginning labs. Using a microscope is a lot of fun. At the end of this course you will be good at it.

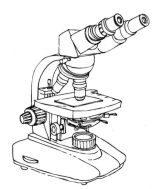

Binocular Microscope

Materials

- Microscope Lab sheet
  (choose the lab sheet for the type of microscope you have)
- Microscope (monocular or binocular)

Procedure

As you read the explanation of what the parts of the microscope are and what each part does, fill in the blanks on the lab sheet naming the part. Optional: You can also write in one or two words telling what the part does.

Instructions are given for both a binocular and a monocular eyepiece. Some sections have different information/instructions for the different eyepieces.

★ **Monocular eyepiece**: a one-eyed eyepiece; you only need to read the monocular portion.

★ **Binocular eyepiece**: a two-eyed eyepiece; you only need to read the binocular portion.

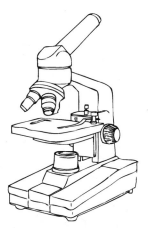

Monocular Microscope

Chapter 1: Microscope Lab 1 *continued*

Parts of the Microscope

Micro is the Greek prefix for "tiny"; *scope* is the Greek root for "to look at." The **microscope** is an instrument that looks at tiny things.

Your Eye

Your eye is not a part of the microscope but we will start with it anyway. People with very good eyesight see things under a microscope more clearly. If you wear glasses, you should wear them when you look through a microscope. It is a good idea to look through the microscope without eye makeup. Small pieces of makeup can fall onto the eyepiece.

★ **Monocular microscope**: It is best to look at images with both eyes open, if you can. If it is confusing for you looking at two separate images, you should practice closing one eye on both sides. Does one side close more easily than the other? If so, make sure you close that eye while you look through the microscope with the other eye.

#1 Eyepiece

The **eyepiece** is the lens at the top of the microscope. It has a magnification of 10x. **Magnification** is the number of times larger that something appears to be when looking at it with that magnification. A lens with a magnification of 10x will make something look 10 times larger.

★ **Monocular microscope**: *Mono* is the Greek prefix meaning "one." *Ocu* is the Latin root for "eye." The monocular microscope has a one-eyed eyepiece.

★ **Binocular microscope**: *Bi* is the Latin prefix meaning "two." The binocular microscope has a two-eyed eyepiece. The amount of space between eyes varies from person to person. The individual eyepieces of a binocular microscope must be adjustable for optimal use.

#2 Tube

The **tube** connects the eyepiece to the nosepiece.

#3 Nosepiece

The **nosepiece** is a flat disk that revolves. It holds the objective lenses.

#4 Objective Lenses

The **objective lenses** are the magnifying lenses that are closest to the object being magnified (the specimen). By themselves, these lenses magnify 4x, 10x, 40x, or 100x (binocular only). Remember the eyepiece had a magnification of 10x. So when the magnification from the eyepiece is multiplied to the magnification from the objective lens, you get 10x that of just the objective lens.

Chapter 1: Microscope Lab 1 *continued*

Scanning	4x	x 10x = 40x magnification
Low power	10x	x 10x = 100x magnification
High dry	40x	x 10x = 400x magnification
Oil immersion	100x	x 10x = 1000x magnification (binocular microscope only)

Each objective lens has important information stamped into the metal.

The 40x lens has this stamp:

4 = Magnification. The object will appear 4 times bigger using this lens alone, but remember it is used with the eyepiece.

0.10 = Numerical aperture (abbreviated NA). The larger the NA, the sharper the object will appear.

160 = Optical tube length. Even though the objective lenses are not the same strength, they all have the same optical tube length of 160mm.

0.17 = Suggested thickness for the slide cover of slides used with this lens. Pay close attention to this number when purchasing slide covers and making up slides.

★ **Monocular microscope**: There are three objective lenses that magnify to: 40x, 100x, 400x.

★ **Binocular microscope**: There are four objective lenses that magnify to: 40x, 100x, 400x, and 1000x. The 1000x lens is an oil immersion (OI) lens. "Oil" is stamped into this lens.

#5 Arm
The *arm* connects the tube to the base.

#6 Stage
The *stage* is where you place slides to look at. There are stage clips on the stage. There are two stage knobs on the stage.

#7 Stage clips (monocular), Slide (binocular)
The *stage clips* hold the *slide* in place on the stage. There can be stage clips on binocular microscopes. Draw them in if your binocular microscope has them.

#8 Stage knobs (binocular)
There are two *stage knobs* on the stage. One moves the stage left to right. The other moves the stage toward and away from you. There can be stage knobs on monocular microscopes. Draw them in if your monocular microscope has them.

Chapter 1: Microscope Lab 1 *continued*

#9 Monocular microscope only: 5-hole rotating disc diaphragm

The *diaphragm* has five different-sized holes. The purpose of the diaphragm is to focus light up from the base through the specimen. Each hole lets through a different amount of light. With light microscopy, more is not always better.

#9 Binocular microscope only: Condenser lens

The *condenser lens* is also called the Abbe condenser, after its inventor. The purpose of the condenser lens is to focus light up from the base through the specimen. There is an arm on the condenser that moves back and forth to adjust the amount of light. With light microscopy, more is not always better.

#10 Coarse focusing adjustment knob and #11 Fine focusing adjustment knob

The *coarse* and *fine focusing adjustment knobs* are nested. They are located on the side of the arm. The larger knob is the coarse focusing adjustment knob. The smaller knob is the fine focusing adjustment knob. The coarse and fine focusing knobs move the stage up and down, closer to and farther away from the objective lens.

#12 Illuminator

The *illuminator* is the light source for a compound microscope.

#13 Base

The *base* supports the microscope and houses the light source.

#14 Compound light microscope

All these parts make up your *compound light microscope*. Compound microscopes have two lenses, the eyepiece and the objective lens, which work together to magnify the specimen. Light microscopes use visible light. Therefore, the *compound light microscope* magnifies through two lenses, the eyepiece and the objective lens, using visible light. The type of compound light microscope used for these experiments is a bright field microscope. Bright field microscopes form a dark image against a more brightly lit background.

Another name you should know: **specimen** or **object**

The *specimen* or *object* is the item being looked at on the slide.

Your Microscope: Parts

Chapter 1: Microscope Lab Sheet

Name_____   **Date**_____

Monocular Microscope

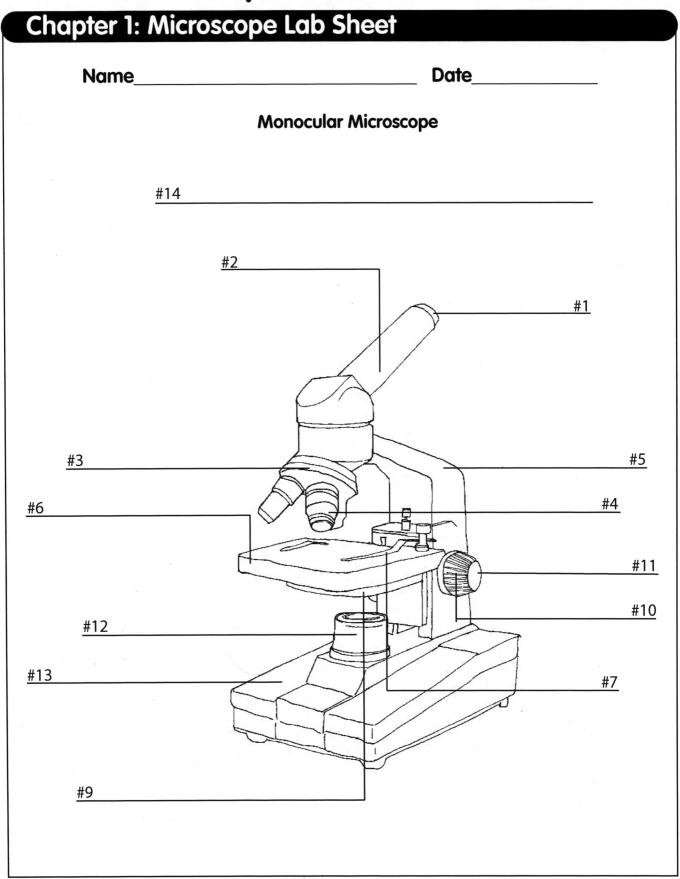

#14 _____

#2

#1

#3   #5

#6   #4

                                                                                                       #11

                                                                                                       #10

#12

#13   #7

#9

Your Microscope: Parts

Chapter 1: Microscope Lab Sheet

Name_____ Date_____

Binocular Microscope

#14 _____

#2

#1

#5

#3

#10

#4

#6

#11

#8

#7

#12

#13

#9

Explore Your Microscope: Focus
Chapter 1: Microscope Lab 2

Materials

- Compound light microscope
- 9cm x 4cm piece of smooth white paper with black type printed on it
- Scissors
- X-Acto knife
- Tape

- Slide
- Catalogue with color pictures
- Microscope view sheet
- Pencil with eraser (just in case)
- Piece of cardboard or a cutting mat

Procedure

1. Rotate the nosepiece so that the lowest power objective lens, 40x, will be focused on the specimen, the piece of paper. This is ALWAYS the starting position when you use your microscope. The lowest-power objective lens is the shortest one with the 4 stamped on it.

2. Put the piece of paper that has type onto the stage, using the stage clips to hold it in place. You will not use the slide for this part of the experiment.

3. Turn the coarse focusing knob until the stage has been moved up as close to the objective lens as it will go.

4. Turn on the light for the microscope.

5. Look through the eyepiece. Do not worry about focusing yet. Are you looking at type when you look through the eyepiece? If not, turn the knobs on the stage until you can see the black type. Find a letter and center it. Because there is no slide, you might need to move the paper with your finger.

   ★ **Monocular microscope** – Use the circle and pointer you see through the eyepiece to help center the paper on a letter.

   ★ **Binocular microscope** – Adjust the eyepieces for your eyes. Use the pointer you see through the eyepiece to help center the paper on a letter.

6. Turn the coarse focusing knob while you are looking through the eyepiece. Stop turning the knob at the clearest point.

Chapter 1: Microscope Lab 2 *continued*

7. Adjust the amount of light.

   ★ **Monocular microscope** – Rotate the disc diaphragm located under the stage to see which of the five holes lets through the optimum amount of light. Hold the microscope with one hand while rotating the disc with the other.

   ★ **Binocular microscope** – Move the arm on the condenser lens to the position that lets through the optimum amount of light. Also, turn the illuminator knob on the base.

8. Turn the fine focusing knob so that the print is very clear. You should see random dots of overspray from the print.

9. Draw what you see through the eyepiece on the microscope view sheet.

10. Rules for drawing specimens:

    a. Use pencil

    b. Write a title

    c. Write the date that you are drawing the specimen

    d. Label each drawing

    e. Make sure the magnification is correctly labeled

    f. Do your best at drawing the specimen the same size as seen in your field of view. The ***field of view*** is what you see when you look through the eyepiece.

11. Turn the nosepiece so the 100x lens is focused on the specimen. Look at the specimen. Do you still see type? When you look at a higher magnification, you are looking at a smaller overall area. Sometimes the part you are looking at will no longer be in the area seen through the lens. If you don't see the type, carefully move the knobs on the stage so the type is seen through the lens.

12. Draw what you see through the eyepiece on the microscope view sheet.

13. Turn the nosepiece so the 400x lens is focused on the specimen. Look at the specimen. Remember, when you look at a higher magnification you are looking at a smaller overall area. Sometimes the part you were looking at will no longer be in the area seen through the lens. Do you still see type? If not, move the stage until you do. What you will also see at this magnification are the fibers making up the paper. You might need to play with the focus to see the paper fibers.

14. Draw the view you see through the eyepiece on the microscope view sheet.

15. Look through the catalogue and find a picture with lots of different colors in it.

16. Cut or tear out this picture.

17. Take the slide and CAREFULLY use the X-Acto knife to cut the picture so that it is the same size as the slide.

Chapter 1: Microscope Lab 2 *continued*

18. Using small pieces of tape, carefully tape the edges of the picture to two ends of the slide. Make sure the picture is flat on the slide. Do not have tape running the length of the slide.

19. Make sure the lowest-power objective lens, the shortest, is focused on the specimen. Put the slide on the stage. Focus and look at the picture with the 40x, 100x, and 400x objective lenses. Play around with the stage knobs to see how they move the slide. Find out how color catalogs are made using only four colors (CMYK = cyan, magenta, yellow, and black). Amazing, isn't it?

20. When you are finished with this lab, remember to
    - Remove the specimen from the stage
    - Turn off the microscope light
    - Cover your microscope
    - Clean off the slide

How to Clean Slides

Slides and slide covers can be cleaned and reused. They are glass and break easily, especially the slide covers, so you need to be careful. To properly clean your slides in this course, you will need: a small dish, dish soap (Dawn is recommended), water, and a place to dry slides.

1. Separate the slide covers and slide.
2. Rinse the specimen and stain off the slide and slide cover.
3. Put a very small amount of dish soap on the slide where the specimen was and rub the slide. Take your soapy fingers and rub the slide cover. If you used the oil immersion lens, make sure you get all the oil off of the slide cover.
4. Rinse off the soap. Make sure you get all the soap off the slide and slide cover.
5. Set them out to dry. Or dry with a soft cloth, especially if you have hard water.
6. If any slides or slide covers break, throw them in the trash.

For More Fun:

Check out the drawings you created on the microscope view sheet. How does ink from a pen look different from ink from a printer with a microscope?

"K" indicates black in the four-color print processing used for your catalogue picture (CMYK). But it doesn't stand for the word "black." Research what it does stand for and why.

Notes:
- The 1000x oil immersion lens on the binocular microscope is not used in the lab.
- Color print uses CMYK processing. These four colors *reflect* or *absorb* light on the page. But your computer screen *emits* only three colors as light—red, green, and blue light (or RGB).

Your Microscope: Focus

Chapter 1: Microscope View Sheet

Name_____ Date_____

Specimen _____

Type of mount_____ Type of stain used_____

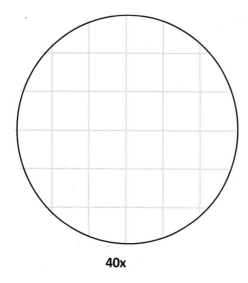

40x

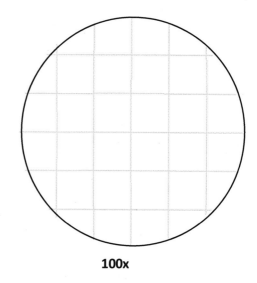

100x

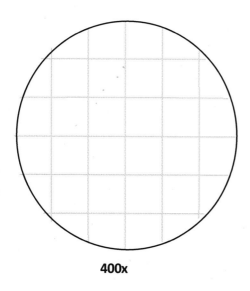

400x

Comments:

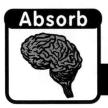

Absorb

Polio

Chapter 1: Famous Science Series

The Famous Science Series is a research assignment. Use your computer or library to find the answers to the questions.

Famous Pathogen: Polio (Poliomyelitis virus)

What is polio? How is it transmitted?

What does it do to a person who is infected with it? What is paralytic polio?

How long has polio been infecting people?

Which U.S. president had polio? When did he serve as president? How old was he when he contracted polio?

Who discovered the polio vaccine?

All Living Things
Chapter 1: Show What You Know

Fill in the blanks

This penguin is a living being. It is a(n)

_____.

Use the nine characteristics of life in the blanks below:

The penguin eats fish. This is how it takes in _____ .

After it eats fish, it has to get rid of _____ .

Laying eggs is part of how the penguin _____ .

Penguins _____ when they swim through the water.

Penguins ruffle up their feathers, trapping warm air near their bodies to help them stay warm. This is one way penguins _____ .

This penguin's blood carries food to its cells and carries waste away from its cells. That is because penguins have _____ .

This penguin is made from many more than one _____ .

Penguins get energy from the food they eat. Penguins have _____ .

A baby penguin _____ after it hatches from the egg on its way to becoming an adult.

Chapter 2: Types

Explore

Death to the Prokaryotes!

Chapter 2: Lab

In 1795, Napoleon Bonaparte needed to feed the French army while they were on the march. In those days, most armies traveled by walking. The French were at war, and men at war doing all that fighting and walking needed to be well fed. A famous quote of Napoleon's is, "An army travels on its stomach." This means a hungry army doesn't do very well, and a well-fed army does. He was concerned about the quality of the food available to his men. Food spoilage was a major problem in those days. There were no refrigerators. Food was kept safe by salting it, drying it, smoking it, or cooking with sugar. Napoleon wanted something healthier. He offered a prize of 12,000 francs to the person who figured out a better method for preserving food.

In 1809, a French chef named Nicolas Appert claimed the prize. Appert had spent ten years experimenting with a method that eliminated air from containers of food. At that time, people thought air caused food to spoil, which is why Appert's method focused on eliminating it. Appert discovered that if you heat jars with a cork tightly sealing the jar to make it an airtight container, the cork will form an airtight seal and the food will not spoil. Appert is considered the "father of canning." If you have ever eaten food that comes out of a jar or a can, you have eaten canned food.

It was not until 50 years later that the famous scientist Louis Pasteur determined the true cause of spoilage. It was not the air. The cause of the spoilage was microorganisms in unsterilized foods. One major type of microorganism that causes spoilage is bacteria, which are prokaryotes.

The experiment you are performing today deals with the area of biology referred to as food safety. Food safety is as important today as it was in Napoleon's day. Unless you eat food fresh from scratch every time you eat, you eat food that has been preserved, most likely by someone you do not know. Every so often, you hear in the news about a food processing plant that has not been careful about following all the steps needed to keep the food they sell safe. Today you are going to learn what some of those steps are.

In this lab, you are going to can applesauce. The process of sterilizing food in containers using heat is called canning, even when you use a jar. You will be using a similar method to that developed by Nicolas Appert. The equipment used has changed quite a bit, but the method hasn't changed much since it was developed over 200 years ago. Microorganisms, which are invisible to the naked eye, are all around you and on the food you eat. Many of them, like those in yogurt, are good for you. Some

Chapter 2: Lab *continued*

Nicolas Appert

are harmful. Using heat to destroy microorganisms kills them. When you kill the organisms that cause spoilage, you stop the spoilage from occurring. You will be killing lots of prokaryotes when you heat the applesauce. When the jars are boiled, air is forced out of the jar and a vacuum seal forms between the lid and the jar. This prevents new microorganisms from entering and contaminating the food. Sterilizing the contents of the jars takes time. It does not take very long to force the air from the jars, but it takes several minutes of boiling to kill certain strains of bacteria. One sample will not go through the sterilization process, so you can see what happens to unsterilized applesauce. Do you think it will spoil? What will happen to the applesauce that is sterilized? Let's experiment and find out.

Materials

- Two ½ pint canning jars
- Two unused rings and lids that fit the jar
- Hot water, for washing the jars and lids before use
- Dish soap
- 1 kg 500g (1½ kg) apples (any variety works). This looks like a lot of apples, but it isn't. DO NOT use bruised, old, or rotting fruit.

- Knife
- ½ cup sugar (optional)
- Apple peeler
- Cutting board
- Tall pot, tall enough for the jars + 8 cm (3 in) of water + some room above that so the water does not boil over
- Lid for pot (optional)
- Pot for cooking applesauce

- Plastic container with lid
- Cooking source
- Wooden spoon
- Food processor or blender for smooth applesauce (optional)
- Potato masher for less chunky applesauce (optional)
- Permanent marker
- Timer

Procedure

1. Wash the jars, rings, and lids with warm, soapy water. Rinse them well and set them aside.

2. Wash and peel the apples. Cut away any bruises you see. Chop the apples, discarding the apple cores with their seeds.

3. Put the apples and sugar (if using) into a pot on your stove. Turn the burner to medium. Set the timer for 20 minutes. Stir occasionally. Do not let the bottom scorch. The heat of stove burners varies; therefore cooking times vary. The apples in the applesauce should be soft when the applesauce is done.

4. Take a small sample of applesauce, let it cool and taste it, and make sure there is enough sugar. Add a little more sugar if necessary. When the apples have softened, turn off the heat. Let the apple mixture cool until it is cool enough for you to spoon into the jars.

Chapter 2: Lab *continued*

5. Use the potato masher to mash the applesauce until it has very few chunks. If you want applesauce that has no lumps, you can put this mixture into a food processor or blender, and process until it is smooth.

6. Fill the two jars with applesauce. It is VERY important you leave headspace in the jar. Headspace is air between the top of the jar and the product being canned. Make sure there is 2 cm, a little less than 1 in, headspace between the top of the jar and the applesauce.

7. Wipe the rim of the jars. Set the lid onto the jar and screw the ring over the top of the lid.

8. Take ¼ cup of the applesauce and put it into the plastic container. Set the cover on the container but do not seal the container. This applesauce will sit loosely covered, until it begins to go bad. Once it starts to spoil, you will be looking at lots of microorganisms. You cannot see just one; there have to be lots for you to see them. Do not eat this now or at any time. There is a range of time this will take to spoil. If you live somewhere warm and humid, the applesauce will spoil quickly. If you live somewhere cool and dry, the applesauce will take longer to spoil.

9. The rest of the applesauce is for you to eat today.

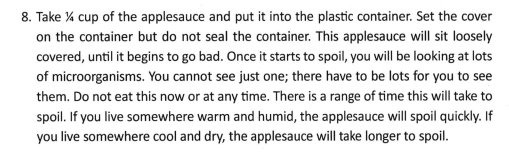

10. Put the two jars into a pot. Fill the pot with water so that the water is 8 cm above the top of the jars. Put a lid on the pot. Turn the heat up to high. When the water boils, set your timer for 30 minutes. You can turn the heat down, but make sure the water continues to boil.

11. After 30 minutes has elapsed, turn off the heat. The jars and water are VERY hot right after the canning process is complete. When the water has cooled, take the jars out of the water. Press lightly on the tops of the lids. The lids should feel tightly sealed.

12. Dry the jars.

13. With a permanent marker, write the date that is two weeks away on one of the jars and the date that is two months away on the other jar. Those are the dates you get to open and eat your applesauce. You have preserved it and killed a whole lot of prokaryotes in the process, so it will be okay to eat.

14. Complete the lab sheet.

15. When you are ready to eat the applesauce, make sure the lids are still vacuum-sealed. If they are not, DO NOT eat the applesauce. That means the applesauce has spoiled.

Preparation for the next lab: Soak a cork in water the night before.

Death to the Prokaryotes!

Chapter 2: Lab Sheet

Name_____ **Date**_____

There are several steps in the canning process that need to be followed to make sure canned food is safe. Explain the danger to food safety at each of the following steps if the proper procedure was not followed.

If old or rotting apples were used . . .
If bruises were not cut from the apples . . .
If the jars, rims, and lids were not clean . . .
If the seal between the jar and rim was not tight . . .
If the applesauce was not cooked as long as it should have been . . .

Discovering Cells

Chapter 2: Microscope Lab

* Lab preparation: Soak the cork in water the night before performing this lab.

Robert Hooke

In 1665, Robert Hooke cut a very thin slice from a piece of cork. Then he put the cork under a compound microscope that he made. Looking through his microscope, he saw tiny boxlike structures that reminded him of monk's cells, the living quarters of monks. That is how cells were discovered and named.

The cork cells that he saw were dead cells. For that reason, he did not see a nucleus, cell membrane, cytoplasm, or genetic material. He saw the cell wall that surrounds plant cells and gives plants their structure, even after the cells have died.

In this experiment, you will take a thin slice of cork and look at it under your microscope. Through the lens of your microscope, you will "discover" cells, just as Robert Hooke did over 300 years ago.

You will make a wet mount slide. Wet mount slides are made with a slide, a specimen, a little bit of water, and a slide cover. The water fills up the space between the slide cover and the slide, this allows light to pass more easily through the prepared slide. Many types of specimen look better with wet mount technique.

Materials

- Compound light microscope
- Slide
- Slide cover (glass is preferred over plastic)
- Optical lens wipes
- Bottle Cork
- X-Acto knife with new blade

- Cutting board
- 1 cc syringe, without the needle (optional, but very handy when making wet mount slides)
- Glass
- Water
- Tweezers with pointed tips, optional

Procedure

How to Prepare a Wet Mount Slide

★ Do not let wet mount slides sit too long after they have been prepared. They will dry out.

★ Make sure your slide and slide cover are clean with no fingerprints or other smudges. Slides and slide covers should be handled on the sides to avoid this.

Chapter 2: Microscope Lab *continued*

1. Take the cork out of water. Slice as thin a slice of cork as you can. Be VERY CAREFUL when making these slices so you do not cut yourself. The slice does not need to be very big around. You are going to look at it under a microscope. The slice I used was .3 cm by .15 cm. Make several slices. You will get better making thin slices with some practice. If the cork is too thick, you will not have a good clear view of the cells. For the best view, the light from the base of the microscope needs to shine through the cork.

2. When you have a cork slice you are satisfied with, put it on the slide.

3. Syringe up about 4 ml of water. One drop at a time, drip water onto the cork on the slide. One to two drops should be enough. Make sure the cork does not float off the slide. Do not put on too much or too little water. The more slides you make the better you will get at judging the amount of water to put on the slide. Be careful to keep the specimen on the slide and under the slide cover with very little to no visible air. Do not worry about it being off center a bit.

4. Put the slide cover on, being careful not to make fingerprints. You want to have the slide cover on the slide with no, or very few, air bubbles. Do not worry about water squishing out of the slide cover at this point. If air bubbles are visible under the slide cover, take the cover off, drip one or two more drops of water onto the slide and put the cover back on. Squeeze all the water out of the syringe into the glass of water. Very carefully, so you do not break it, press on the slide cover with the end of the syringe to try to squeeze the air out. If this doesn't work, you may need to shave a thinner slice of cork. Either that, or try thinning the slice you are working with.

5. If water has leaked out from under the slide cover, put the end of a paper towel on the water so that the paper towel passively soaks up the water. Do not let it soak up all the water from under the slide cover, though. Make sure there is no water on the bottom of your slide.

Viewing the Slide

★ Warning: You need to be very careful when rotating and focusing the higher-magnification objective lenses. The objective lenses can break the slides if they are focused down too hard; even worse, you could damage the objective lens. Just pay attention and always start with the 40x objective lens when you are focusing on a specimen.

6. Rotate the nosepiece so that the lowest-power objective lens, 40x, is focused on the stage.

7. Put the cork slide onto the stage using the stage clips to hold it in place. To do this: Pull back on the lever on the right stage clip, then put the slide resting against the bottom corner of the left side stage clip. With your finger, carefully let the right side stage clip close to hold the slide in place.

8. Turn on the microscope light. Use the knobs on the stage to move the specimen into the circle of light coming from the illuminator up through the specimen. Take your eyes away from the microscope and slowly move the slide with the knobs on the stage. Notice that when you do this the slide moves in the opposite direction you would expect. This is called ***microscopic inversion***.

Chapter 2: Microscope Lab *continued*

9. Look through the eyepiece:

   A. Center the specimen.

   B. Turn the coarse focusing knob while you are looking through the eyepiece. Stop turning the knob at the clearest point for this knob.

   C. Adjust the amount of light using either the diaphragm or the condenser lens.

   D. Refocus with the coarse focusing knob.

   E. Focus with the fine focusing knob.

10. You should see the cork cells clearly. If you have a very small specimen, you might see all of it. Look over the specimen. Find the part that looks like the thinnest slice; this is where the most light is getting through the sample. You might need to move the stage knobs to move the cork. Check for air bubbles. Move away from air bubbles; they interfere with your view. If the thinnest part of the cork is also the part with air bubbles, you might want to work with the slide some more. Possibly make a thinner slice from the cork or take off the cover and add more water.

11. Draw what you see through the eyepiece on your microscope view sheet. Your view will have many similar-looking cells:

    • Draw the entire outline of the specimen, if the specimen does not take up the entire field of view.

    • Draw a few of the cells.

    • Note on your view sheet that the rest of the inside of the specimen looks like what you have already drawn.

12. Turn the nosepiece so the 100x lens is focused on the specimen. Look at the specimen. Do you still see cork? When you look at a higher magnification, you are looking at a smaller overall area. Sometimes the part you are looking at will no longer be in the area seen through the lens. If you don't see the cork, move the stage knobs so the lens is centered on the cork. Use the focusing knobs so that the cork is at its clearest.

13. Draw the view through the eyepiece on your lab sheet. Use the same strategy you used at 40x.

14. Turn the nosepiece to the 400x lens and repeat the above procedure.

Finishing Up

15. When you are finished with this lab, remember to

    • Remove the specimen from the stage

    • Turn off the microscope light

    • Cover your microscope

    • Clean off the slide and slide cover

Discovering Cells

Chapter 2: Microscope Lab Sheet

Name_____ **Date**_____

Specimen _____

Type of mount_____ Type of stain used_____

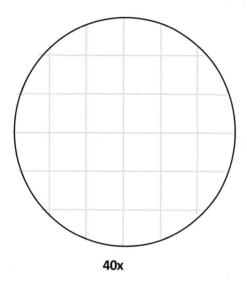

40x

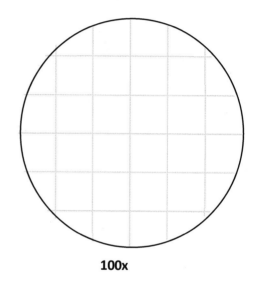

100x

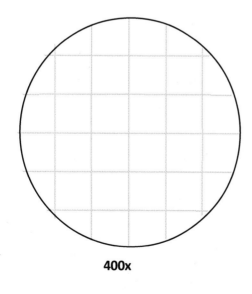

400x

Comments:

Pandia PRESS

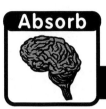 Absorb Antonie van Leeuwenhoek
Chapter 2: Famous Science Series

Father of Microbiology: Antonie van Leeuwenhoek [LAYU-wen-hook]

Why is Antonie van Leeuwenhoek famous? What did he discover? What did he use to discover them?

Antonie van Leeuwenhoek

When and where was he born?

When did he die?

He was inspired after reading a famous book written by Robert Hooke. What is the title?

It has been speculated that the Dutch painter Johannes Vermeer used optical aids produced by van Leeuwenhoek. How would these have helped Vermeer?

How many microscopes did van Leeuwenhoek make? What happened to them?

Types
Chapter 2: Show What You Know

Multiple Choice

1. A shark is made from

   ○ prokaryotic cells
   ○ eukaryotic cells
   ○ prokaryotic and eukaryotic cells
   ○ I need more information

2. The bacteria that causes strep throat are made from

   ○ prokaryotic cells
   ○ eukaryotic cells
   ○ prokaryotic and eukaryotic cells
   ○ I need more information

3. The basic unit of structure and function of an organism is called

   ○ a prokaryote
   ○ an eukaryote
   ○ a cell
   ○ an amoeba

4. Unicellular organisms are

   ○ all prokaryotes
   ○ all eukaryotes
   ○ prokaryotes and eukaryotes

Pandia PRESS

Chapter 2: Show What You Know *continued*

Draw

5. Draw and label the cell membrane, the cytoplasm, and the genetic material. Draw the nucleus, where applicable.

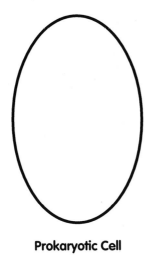

Prokaryotic Cell

Eukaryotic Cell

Fill in the Blanks

The _____ theory states . . .

6. Every _____ is made of one or more _____ .

7. _____ come only from other living _____ .

8. _____ are the basic unit of _____ and

_____ needed to support _____ .

Question

9. What famous scientist coined the term *cell*? Why didn't he see a nucleus, cell membrane, cytoplasm, or genetic material?

Chapter 2: Show What You Know *continued*

10. Match the word with the best definition.

Unicellular ⬭   ⬭ many-celled

Cell ⬭   ⬭ a jelly-like material inside all cells

Cell membrane ⬭   ⬭ genetic material, deoxyribonucleic acid

Cytoplasm ⬭   ⬭ one-celled

DNA ⬭   ⬭ an organism whose DNA is located in the cytoplasm

Eukaryote ⬭   ⬭ the basic unit of life

Multicellular ⬭   ⬭ an organism whose DNA is located inside the nucleus

Prokaryote ⬭   ⬭ encloses and protects the inside of the cell

Pandia PRESS

Chapter 3: The Inside Story

Explore

Cells Are Three-Dimensional

Chapter 3: Lab

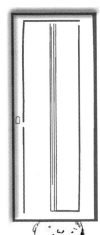

Flat Stanley is two-dimensional. He slides under the door.

Imagine this scenario: Let's say you come home, walk through your front door, and walk into your bedroom. The distance you walked from your front door to your bedroom is one-dimensional. You might turn corners and even cross back over your path. None of that changes the fact that the total distance you walked is in one dimension: length.

One-Dimensional has a measurement of one dimension
Length = Distance

When you walk into your room, you see that your two-year-old sister has poured paint all over your carpet. Oh yea. Your bedroom now needs new carpet. To determine the amount of carpet to buy for your bedroom you need to know two dimensions: length and width. This measurement is two-dimensional. It is a measurement of area.

Two-Dimensional has a measurement of two dimensions
Length x Width = Area

When you look at the drawing of a cell on a page of paper, it looks two-dimensional. They may look that way on paper, but cells are not two-dimensional. If they were, you would be like Flat Stanley, from the *Flat Stanley* books. You would have a length and a width, but you would not have any depth. When you went into a room you would not have to open the door; you could slide right under it.

Let's get back to our scenario. Your parents tell you to forget the carpet. They want you to turn your bedroom into an indoor swimming pool instead. Oh yea! How much water will your new indoor bedroom pool hold? To determine the amount of water your bedroom will hold, you need the length of your bedroom, the width of your bedroom, and the height, or depth, of your bedroom. This measurement is three-dimensional. It is a measurement of volume—measurement of how much water would fill the space of your room.

Three-Dimensional has a measurement of three dimensions
Length x Width x Height (Depth) = Volume

Chapter 3: Lab *continued*

Cells have a length, a width, and a depth. Therefore, cells and organisms made from cells, like you, are three-dimensional. Today you are going to make a three-dimensional model of an animal cell. You are going to make a cellular model that has been cut in half, so you can see the three-dimensional organelles that are inside.

The model you make will be of a typical animal cell. A typical cell is a representative of a cell. You do not have a type of cell in your body called *typical*. The model shows the three-dimensional structure of cells and organelles, their relative size to each other, and the approximate shape of the organelles.

Materials

- White glue that dries clear
- Super glue
- Plastic wrap
- Knife
- Scissors
- Toothpick
- Ruler
- Organelle Labels lab sheet
- Tape

- Pins or toothpicks, for labels
- Modeling clay (e.g., Sculpey) in the following colors: blue, orange, red, yellow, pink, purple, brown. (You can substitute colors, but do not use green. Green is the color of chloroplasts found in plant cells.)
- Cookie sheet
- Oven

- 15 cm (6 in) bowl
- 3 cups Plaster of Paris
- Container for mixing Plaster of Paris
- Something with which to stir Plaster of Paris
- Measuring cup
- 2 cups water

Procedure

★ Do not make Plaster of Paris before instructed. It dries VERY quickly.

★ There are instructions for putting labels in the cell using pins or toothpicks and labeled strips of paper. When this is done, the cell can easily be referenced as to the organelle and its function. The problem is it makes the cell busy-looking. An option is to fill in the label sheet, color-code it so title and description match the color of the organelle, and then keep it for reference. Choose whichever labeling method you prefer.

1. Make the organelles. Roll the Sculpey into the shapes described on the next page. Read the entire How to Make Organelles before beginning. Use your ruler to get the measurements correct. The measurements do not need to be exact, but should be close. As each organelle is molded, set it on the cookie sheet.

Chapter 3: Lab *continued*

How to Make Organelles

Below is a list of the organelles you need to make.

Organelle	Clay Color	Shape of Organelle	Size of Organelle	# to Make	Examples (not to scale)
Nucleus	Blue	Circle	4½ cm	1	
Mitochondria	Red	Oval	2 cm x 1 cm	3	
DNA	Orange	Very thin rope	8 cm long	5	
Rough Endoplasmic Reticulum	Yellow	Flat rectangular disk	4½ cm x 4 cm 3½ cm x 3 cm 3 cm x 3 cm 2 cm x 2 cm	1 1 1 1 = 4 total	
Smooth Endoplasmic Reticulum	Yellow	Tube shape	1½ cm x 1 cm	6	
Golgi Apparatus	Pink	Flat pancake-shaped disks	Diameter = 3 cm Diameter = 2 cm	1 2 = 3 total	
Ribosomes	Purple	Circle	0.3 cm	36	
Vacuoles	Brown	Circle	Between ½ and 1 cm	6	

Chapter 3: Lab *continued*

2. Make the nucleus as a sphere with a diameter of 4½ cm. Then cut ¼ of the circle away. Think of a clock at 3 o'clock. Cut the clay away from in between the minute hand at 12 and the hour hand at 3.

3. Use a toothpick to poke about 15 holes on each side in the two inside cuts of the nucleus.

4. Take the DNA ropes and put them in the cut-away section of the nucleus. Gently, use the toothpick to anchor the DNA into the nucleus. You want the DNA to become a molded part of the nucleus. The DNA should look sort of like a rounded mass of spaghetti. It should maintain the rounded shape of the nucleus that was cut away. You cut the ¼ out of the nucleus, so you could peek inside to see the DNA, not so that the DNA could escape from the nucleus. Put this on the cookie sheet.

5. Take the disks that are the rough endoplasmic reticulum and gently cup them to the rounded shape of the nucleus. When the cell is put together, these nest next to the nucleus. Put these on the cookie sheet.

6. Finish making all the rest of the organelles and place them on the cookie sheet. If using Sculpey clay, bake them at 275°F for the following time:
   - 10 minutes for the ribosomes, Golgi apparatus, and rough endoplasmic reticulum
   - 5 minutes for the vacuoles and smooth endoplasmic reticulum
   - 25 minutes for the nucleus

7. While the organelles are baking, write the functions of each type of organelle on the labels on the lab sheet. Optional: Make a copy of the lab sheet, cut out the labels, and tape them onto toothpicks or pins, creating label flags to use in step #18.

8. When the clay has cooled, put the rough endoplasmic reticulum together:

The rough endoplasmic reticulum is going to be pushed into the Plaster of Paris. You are going to glue ribosomes on both flat sides only on the TOP half of each disk. This makes it easier to put it into the Plaster of Paris. In a real cell, though, the rough endoplasmic reticulum has ribosomes across its entire surface. Take 26 ribosomes and, with super glue, glue half of them on the top half of one side of each yellow disk. Put a dot of super glue on the rough endoplasmic reticulum and RIGHT AWAY put a ribosome on it. Let the glue dry then turn the disks over and glue the other half on the top half of the other side.

9. Lay out the organelles to get an idea how everything is going to fit.

10. Cut a sheet of plastic wrap that is larger than the inside of the bowl when it is in the bowl. Lay the plastic wrap in the bowl with the ends overhanging. You will use the ends to help get the dried cell out of the bowl.

11. Make the Plaster of Paris. Gently stir 2 cups of water into 3 cups of Plaster of Paris.

Chapter 3: Lab *continued*

12. Pour it into the bowl about an inch from the top of the bowl. Be careful that the plastic wrap is above the level of the Plaster of Paris. Check the Plaster of Paris every ONE minute to see if it has begun to set. It should only take about five minutes before you can start putting the organelles into the Plaster of Paris. It might be even less than five minutes. You need to work quickly. You want the plaster to be set enough to hold the organelles, but not too stiff.

13. It is time to put the organelles in the cell. DO NOT randomly place the organelles in the cell. Use the diagram. Start with the nucleus. Push the nucleus part way into the Plaster of Paris. Make sure the DNA is up so it is easy to see.

14. Put the rough endoplasmic reticulum in next. They go right next to the nucleus. The top half with the ribosomes should be sticking out of the Plaster of Paris.

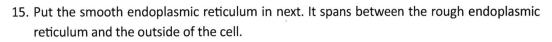

15. Put the smooth endoplasmic reticulum in next. It spans between the rough endoplasmic reticulum and the outside of the cell.

16. Put the Golgi apparatus in next.

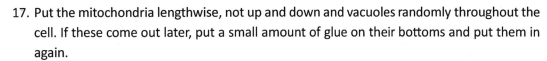

17. Put the mitochondria lengthwise, not up and down and vacuoles randomly throughout the cell. If these come out later, put a small amount of glue on their bottoms and put them in again.

18. Last, spread the ten remaining "free-floating" ribosomes throughout the cell; it should be random. If you need to, use glue to secure them onto the top. Optional: When you are done putting all the organelles in their permanent places, stick the label flags you created in step #7 next to the organelles.

19. Let your cell sit for an hour so the Plaster of Paris can set. Then carefully take the ends of the plastic wrap and lift the cell out of the bowl. The Plaster of Paris covered by the plastic wrap will dry. It may take one to a few days, depending on the humidity where you are.

20. When the Plaster of Paris has dried, carefully squirt white glue on the top of your cell. The glue represents the cytoplasm. Squirt small amounts of glue around the organelles to help keep them in place. Be careful with the glue. You need to coat the top of the cell, but not your work surface. Let the glue dry overnight; it should dry clear.

21. The next day, cut the edges of the plastic wrap so they are flush with the cut edge of the cell. The plastic wrap represents the cell membrane. If it comes off, glue it back onto the outside of the cell.

Cells Are Three-Dimensional

Chapter 3: Lab Sheet

Name_____ Date_____

Organelle Labels

Nucleus	DNA
Smooth Endoplasmic Reticulum	Mitochondria
Golgi Apparatus	Rough Endoplasmic Reticulum
Vacuoles	Free-Floating Ribosomes
Ribosomes	Cell Membrane
Cytoplasm	

Looking at Cells

Chapter 3: Microscope Lab

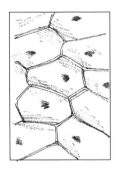

Biologists use stain to see transparent organelles of a cell.

★ The stain used today stains everything: your clothes, your hands, your furniture. Be careful when using it.

★ Now that you know the basics of operating a microscope, there are fewer instructions.

In this lab, you will be comparing animal and plant cells. You will get the animal cells by scraping the inside of your cheek. Don't worry, it won't hurt. The cells on the inside of your cheeks come off easily. The plant cells will come from an onion. The cells today are living cells. At least they were before you put them on slides. You will be able to see cell membranes, nuclei, and cytoplasm in both cells. The onion cells will also have a visible cell wall. The cell wall can make the cell membrane hard to see, but it is there. There are other organelles present and you will see them, but they are too small, even through a microscope lens, to be positively identified.

The organelles of cells can be transparent under the microscope. This makes them difficult to see. Biologists solve this problem by staining them. Staining cells means that dye or stain is put on the cells. The cells' organelles absorb the stain, making them easier to see under the microscope. There are different types of microscope stain. The different stains stain different types of molecules. The type of stain to use depends on what you want to view. Methylene blue is a good stain for looking at a cell's nucleus.

Materials

- Compound light microscope
- 2 slides
- 2 slide covers
- Plastic teaspoon
- Yellow onion
- X-Acto knife
- Paper towel

- Methylene blue
- Syringe
- Water
- Oil, only if you have an oil immersion lens
- Cleaner for the oil, only if you have an oil immersion lens

Chapter 3: Microscope Lab *continued*

Procedure

Viewing an animal cell:

1. Swish the inside of your mouth with water. This gets rid of any food debris you might have in your mouth. Your mouth needs to be clean so you get cells, not food molecules.

2. Drip one drop of water onto a slide and set it down. VERY GENTLY, scrape the inside of your mouth with the end of the plastic teaspoon. Swirl the area of the teaspoon you used to scrape your mouth gently in the water on the slide.

3. Drip one drop of methylene blue stain onto the slide, on top of the cheek cell + water mix. Let it sit for about 30 seconds. Put the slide cover on this.

4. The nucleus is now stained. The problem with viewing it is the cell is in a puddle of stain. You need to get some of the stain off the slide, without losing the stained cells.
   - Leave the slide cover on.
   - Drip a couple of drops of water onto the slide at the edge of the slide cover.
   - Gently place the end of a paper towel alongside the slide cover on the other side from where you dripped the water. Try not to move the slide cover when you do this. Hold the paper towel there until most of the stain is gone from under the slide cover. Make sure there is still some water under the slide cover.
   - Wipe any water or stain that is on the underside of the slide. You do not want to get your stage dirty.

5. Put the slide on the stage. Focus the 40x objective lens. After the slide is in focus, go in search of cheek cells. Focus clearly with the pointer pointed directly onto a cell that gives a good view. Some of the cells will be folded over or in a clump of cells; you do not want to use these cells. Play around with this, maybe even redoing the slide, until you get a good cell to view.

6. Look at the cell with the 100x and the 400x objective lenses. Make sure the pointer is still pointed onto the cell. Choose the view sheet that matches your microscope.
   - Monocular and Binocular: Draw the view as seen through the 400x objective lens. This is the best view for seeing the whole cell.
   - Binocular. For oil immersion lens only: Lower the stage as low as it will go. CAREFULLY drip one drop of the oil onto the top of the slide cover at about the place the lens is focused. Do not move the slide! Raise the stage and focus the 1000x, the oil immersion, objective lens. Draw the view of the nucleus as seen through the 1000x objective lens. You may need to go back to the 100x lens and refocus with the pointer on the nucleus and then come back to the 1000x lens. Do not use the 400x lens. You might get oil on that lens because it can touch the slide cover. You can ruin the lenses not intended for oil immersion by getting oil on them. When you are done, wipe the oil immersion lens with cleaner.

Chapter 3: Microscope Lab *continued*

Viewing a plant cell:

7. Using the X-Acto knife, cut a 1 cm by 6 cm slice of the onion. The depth will be about four layers deep, not counting the outside skin. Peel the layers apart, until you find a piece of the inner surface that is attached to each thick layer. The inner surface looks like tissue paper. Cut a ½ to 1 cm square of the inner surface and lay this out flat on a dry slide. It might help to scrape it flat by gently using the X-Acto knife blade. Drip one drop of methylene blue onto the onion. Let it sit for about 30 seconds.

8. Put the slide cover on and follow steps 4 through 6 above to observe and draw the onion cells as you did the cheek cells.

Looking at Cells

Chapter 3: Microscope View Sheet

Name_____   Date_____

Monocular Microscope

Specimen _____

Type of mount_____   Type of stain used_____

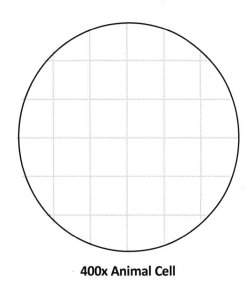

400x Animal Cell

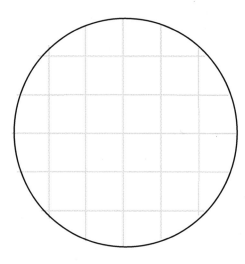

400x Plant Cell

Observations:

You had to be more careful working with the animal cells than the plant cells. Why?

Pandia PRESS

Looking at Cells

Chapter 3: Microscope View Sheet

Name_____ **Date**_____

Binocular Microscope

Specimen _____

Type of mount_____ Type of stain used_____

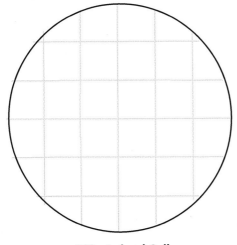

400x Animal Cell

400x Plant Cell

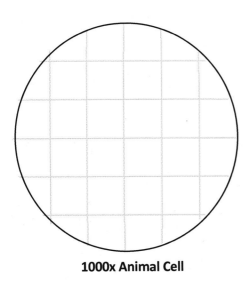

1000x Animal Cell

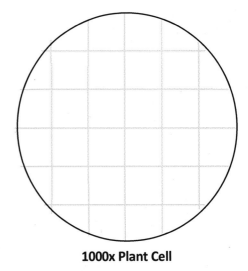

1000x Plant Cell

You had to be much more careful working with the animal cells than the plant cells. Why?

Sickle Cell Anemia

Chapter 3: Famous Science Series

Sickle cell anemia is a disorder affecting some people. Sickle cell anemia causes red blood cells to be abnormally shaped. Draw a picture of a normally shaped red blood cell and a picture of a blood cell from a person with sickle cell anemia.

Normal Red Blood Cell   **Affected Red Blood Cell**

How do people get sickle cell anemia?

Why is it a problem when blood cells are not normally shaped?

What health problems do people with sickle cell anemia have?

Is there a cure for sickle cell anemia?

The Inside Story

Chapter 3: Show What You Know

1. Match the organelle with the phrase describing it.

Nucleus ◯   ◯ Would you like some lipids? I just made some.

Mitochondria ◯   ◯ That tree would be a floppy sack without me.

Vacuole ◯   ◯ I like being the boss.

Ribosomes ◯   ◯ Some people call me the # 3 man of protein synthesis.
                              I make things bigger and better.

Smooth Endoplasmic ◯   ◯ I look like a peppercorn and I make proteins.
Reticulum

Cell Membrane ◯   ◯ Don't call me "trash can"; I prefer "pantry."

Golgi Apparatus ◯   ◯ Green solar panels, that's me.

Rough Endoplasmic ◯   ◯ All cells have one of me. I am a must-have for
Reticulum   outerwear fashion.

Cell Wall ◯   ◯ Just call me the energy maker.

Chloroplasts ◯   ◯ I only look rough because of the ribosomes.
                              I am actually very soft inside.

Chapter 3: Show What You Know *continued*

2. True or False. Below is a list of statements about cell specialization. Write true or false. If false, rewrite the statement making it true.

Unicellular organisms have specialized cells. True or false?

All body cells in an organism have the same DNA, no matter what their specialization.
True or false?

Specialized cells in an organism have different shapes depending on their specialization.
True or false?

All the cells in a multicellular organism have the same number and type of organelles.
True or false?

One cell from a multicellular organism can survive on its own. True or false?

Chapter 3: Show What You Know *continued*

3. Label and Color. Below is a cross-section drawing of a plant cell. Label and color the organelles. Use the organelle names from problem #1 and your Textbook if you need some help remembering. Optional: Cut around the cell and your labels and make a plant cell poster. Place it on a green background.

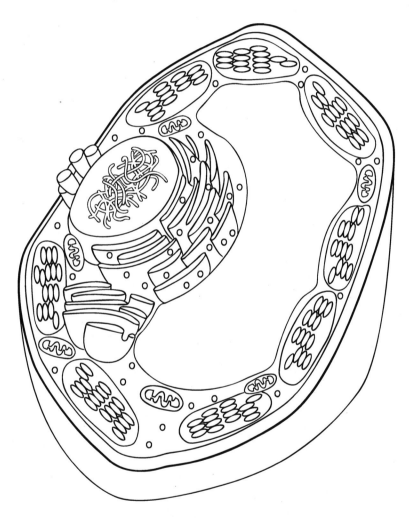

Plant Cell

Chapter 4: The Chemistry of Biology

Explore

My Food Choices
Chapter 4: Lab

The food you eat is full of carbohydrate, lipid, protein, and nucleic acid molecules. It has vitamins and minerals in it too. These are the molecules you need to build cells. Molecules are really small. It would be hard to figure out the right amount of the molecules you should eat every day.

To help, scientists developed dietary guidelines that tell the best foods and amounts to eat to get the different molecules a body needs. This helps people make the right choices about the food they eat: how many servings of each food type and the size the portions should be. Most dietary guidelines divide food into five main groups. To get all the vitamins and minerals your body needs, you must eat foods from all five food groups.

I am growing. I wonder if I have all the molecules I need to make my new cells.

Five Main Food Groups

Food Group	Types of Molecules in the Group
Grain	Carbohydrates Proteins Nucleic Acids
Protein	Proteins Lipids Nucleic Acids
Dairy	Proteins Carbohydrates Lipids (not present in non-fat dairy)
Fruits	Carbohydrates
Vegetables	Carbohydrates Lipids* * Lipids are not found in most vegetables. When they are, they are usually in the form of unsaturated fats.

Chapter 4: Lab *continued*

There are eight amino acids your body cannot make. They must come from the food you eat.

Butter is a solid, saturated, fat. You should not eat too much of this kind of fat. Olive oil is a liquid, unsaturated, fat. This is the best kind of fat for your body.

Grains: Grains come from plants. They are an important source of the carbohydrates your cells need to make energy. Grains are divided into two groups: *whole grains* and *refined grains*. The carbohydrates in whole grains are called complex carbohydrates. Think of it this way: Because there are more parts to the grain, the chemical reactions used to break the carbohydrates into molecules are more complex than those used for refined grains. These more complex chemical reactions give a more steady release of energy. The grains you eat should be whole grains. Refined grains should be eaten sparingly.

Proteins: Meat, beans, seeds, nuts, and eggs are all good sources of protein. Your body needs twenty different types of amino acids for making protein molecules. Your body makes twelve types of amino acids. There are eight more types of amino acids that your body needs but cannot make. These eight amino acids are called *essential amino acids*. It is essential you eat protein with these eight amino acids in them. To get the essential amino acids you need, eat proteins from a variety of sources.

Dairy: Cheese, butter, yogurt, cream, and cream products are all dairy products made from milk. Dairy products are a good source of proteins and the mineral calcium. Dairy products can also be a source of saturated fats, which can be harmful for you. When choosing dairy products, consider non-fat varieties. These have had the saturated fat removed. Left behind is the protein and calcium your cells need.

Fruits and Vegetables: Fresh fruit and vegetables are a source of carbohydrates, vitamins, minerals, and fiber. Like grains, fruits and vegetables are best for you before they are highly processed. Turning fruit into fruit juice removes the complex carbohydrates, many of the vitamins and minerals, and the fiber, all of which are essential to a healthy diet.

When choosing fruits and vegetables, pay attention to color. You should eat a variety of different colors of fruits and vegetables. Fruits and vegetables are made from molecules. The color of fruits and vegetables depends on what molecules are present. The molecules that make a cherry red are different from the molecules that make cantaloupe orange or broccoli green. These different color molecules give your body more choices when it is building cells and fighting diseases.

Oils: Not included in the five main food groups are oils. Oils are a source of lipids. Your body makes some, but not all, of the lipids it needs in the smooth endoplasmic reticulum. There are two types of fats: saturated and unsaturated. Saturated fats, like the fat you see on meat, are solid at room temperature. Unsaturated fats, like olive oil, are liquid at room temperature.

Chapter 4: Lab *continued*

Water: Some food guidelines do not include water. Remember, chemical reactions in your body happen in water. You must make sure you drink enough water every day so your cells have enough to do all the things they need to do.

Today you will research healthy eating guidelines and make a menu for yourself. The next day you will eat from the menu you made. Are you a junk-food junkie, or a healthy eater? If you do not know the answer, this lab will help you find out. If you normally do not eat enough from a category or you eat too much from a category, you should think about making some changes to your diet. Remember, your cells are counting on you to make the right food choices.

Materials

- Internet access
- Food items on your menu
- Calorie counter tool (optional)

Procedure

MyPlate is the nutrition guidelines published by the USDA.

1. Research healthy dietary guidelines online. Good places to look are the USDA's MyPlate Plan, the Mayo Clinic, and the American Heart Association. Avoid fad weight loss diets and those that recommend omitting entire food groups. Visit www.pandiapress.com/weblinks-biology2 for links to helpful websites.

2. As you're researching, take notes regarding how much of each food group you should be eating each day and the recommended portion sizes. For example, the USDA recommends making half of your plate fruits and vegetables. Optionally, you can use an online tool or app to calculate your ideal daily caloric intake, and a calorie counter to determine the calories of your meal plan.

3. Now it is time to make your meal plan. Create a menu for one day on the provided lab sheet. Be sure to choose foods from every food group and pay attention to serving size. There are three spaces for snacks, but you do not have to have three snacks. This is your menu, so tailor it to your eating habits.

   Dietary notes:
   - If you are lactose intolerant or vegan, you will need to get your calcium and vitamin D from a source other than the dairy category.
   - Try to use whole grain for at least half of your grain servings. Popcorn is a whole grain.

4. The next day, use the menu you created when eating. You might need to go shopping for your items. At the end of the day, answer the questions at the bottom of the lab sheet. If you enjoy this activity, make a week's worth of menus.

My Food Choices

Chapter 4: Lab Sheet

Name_____ Date_____

Make a Meal Plan

Breakfast
Snack
Lunch
Snack
Dinner
Snack

How well did you stick to your plan?

Is this close to your normal diet?

From which food group did you eat the most?

From which group did you eat the least?

Pandia PRESS

Nutrition Labels
Chapter 4: Activity

Have you noticed that all food nutrition labels look pretty much the same? That is because in the United States, food nutrition labels have been regulated to be consistent by the FDA (Food and Drug Administration) and the USDA (United States Department of Agriculture) since the 1990s. But have you ever closely examined a food label? Maybe you've glanced at one to learn the number of calories or the grams of fat in the food you're eating. But there's a lot more to food labels than fat and calories. Would you be surprised to learn that most adults (as many as 70 percent) do not know how to properly interpret the nutritional information on a food label*? Today you are going to learn your way around a food label so well that you might be able to teach the adults in your life something about nutrition. Below is the nutrition information from a can of black beans with an explanation of each section of the label.

*According to the U.S. Department of Agriculture's Economic Research Service

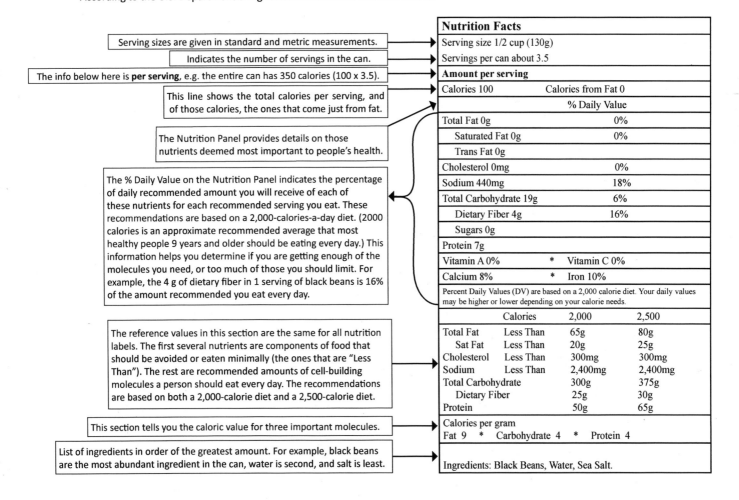

Serving sizes are given in standard and metric measurements.

Indicates the number of servings in the can.

The info below here is **per serving**, e.g. the entire can has 350 calories (100 x 3.5).

This line shows the total calories per serving, and of those calories, the ones that come just from fat.

The Nutrition Panel provides details on those nutrients deemed most important to people's health.

The % Daily Value on the Nutrition Panel indicates the percentage of daily recommended amount you will receive of each of these nutrients for each recommended serving you eat. These recommendations are based on a 2,000-calories-a-day diet. (2000 calories is an approximate recommended average that most healthy people 9 years and older should be eating every day.) This information helps you determine if you are getting enough of the molecules you need, or too much of those you should limit. For example, the 4 g of dietary fiber in 1 serving of black beans is 16% of the amount recommended you eat every day.

The reference values in this section are the same for all nutrition labels. The first several nutrients are components of food that should be avoided or eaten minimally (the ones that are "Less Than"). The rest are recommended amounts of cell-building molecules a person should eat every day. The recommendations are based on both a 2,000-calorie diet and a 2,500-calorie diet.

This section tells you the caloric value for three important molecules.

List of ingredients in order of the greatest amount. For example, black beans are the most abundant ingredient in the can, water is second, and salt is least.

Nutrition Facts

Nutrition Facts		
Serving size 1/2 cup (130g)		
Servings per can about 3.5		
Amount per serving		
Calories 100	Calories from Fat 0	
	% Daily Value	
Total Fat 0g		0%
Saturated Fat 0g		0%
Trans Fat 0g		
Cholesterol 0mg		0%
Sodium 440mg		18%
Total Carbohydrate 19g		6%
Dietary Fiber 4g		16%
Sugars 0g		
Protein 7g		
Vitamin A 0%	*	Vitamin C 0%
Calcium 8%	*	Iron 10%

Percent Daily Values (DV) are based on a 2,000 calorie diet. Your daily values may be higher or lower depending on your calorie needs.

	Calories	2,000	2,500
Total Fat	Less Than	65g	80g
Sat Fat	Less Than	20g	25g
Cholesterol	Less Than	300mg	300mg
Sodium	Less Than	2,400mg	2,400mg
Total Carbohydrate		300g	375g
Dietary Fiber		25g	30g
Protein		50g	65g

Calories per gram
Fat 9 * Carbohydrate 4 * Protein 4

Ingredients: Black Beans, Water, Sea Salt.

Chapter 4: Activity *continued*

Let's take a closer look where most people start (and often stop) on a food label—the calories.

You can see that a single serving of beans (½ cup) has 100 calories.

But just knowing the *number* of calories is not enough. All calories are not equal as far as nutritional value is concerned. It is important to know *where* those calories come from. You can calculate the number of calories per serving coming from the carbohydrates, lipids (fat), and proteins in the food using the information located at the bottom of the label in a section titled "Calories per gram." This information is the same on all food labels. It tells you how many calories a gram of fat, carbohydrate, or protein contains.

Calories from fat is calculated by multiplying the grams of fat found in a serving size of beans (0) by the calories that are normally found in a gram of fat (9).

0 x 9 = 0 calories from fat

The same can be done for calculating calories from carbohydrate and protein:

Carbohydrate: **19 x 4 = 76 calories from carbohydrates**

Protein: **7 x 4 = 28 calories from protein**

Total calories: **0 + 76 + 28 = 104 calories or 100 calories** (It is common to round off when calculating calories per serving.)

Too much fat, cholesterol, sodium, and sugars can lead to health problems. They are listed on the label to help you avoid eating too much of them. Unsaturated fat is not given in the label information. The difference between the total fat and the saturated fat gives the amount of unsaturated fat. Use what you have learned about nutrition labels to answer the questions on the following worksheet. Then on the second page of the worksheet, complete the chart and teach someone else about food labels.

Nutrition Facts

Serving size 1/2 cup (130g)

Servings per bag about 3.5

Amount per serving

Calories 100	Calories from Fat 0	
	% Daily Value	
Total Fat 0g		0%
Saturated Fat 0g		0%
Trans Fat 0g		
Cholesterol 0mg		0%
Sodium 440mg		18%
Total Carbohydrate 19g		6%
Dietary Fiber 4g		16%
Sugars 0g		
Protein 7g		
Vitamin A 0%	*	Vitamin C 0%
Calcium 8%	*	Iron 10%

Percent Daily Values (DV) are based on a 2,000 calorie diet. Your daily values may be higher or lower depending on your calorie needs.

	Calories	2,000	2,500
Total Fat	Less Than	65g	80g
Sat Fat	Less Than	20g	25g
Cholesterol	Less Than	300mg	300mg
Sodium	Less Than	2,400mg	2,400mg
Total Carbohydrate		300g	375g
Dietary Fiber		25g	30g
Protein		50g	65g

Calories per gram

Fat 9   *   Carbohydrate 4   *   Protein 4

Ingredients: Black Beans, Water, Sea Salt.

Nutritional Labels

Chapter 4: Activity Worksheet

Name_____ **Date**_____

Label #1: Whole Wheat Flour	Label #2: Eggs

Nutrition Facts	
Serving size 1/4 cup (30g)	
Servings per container about 76	
Amount per serving	
Calories 110	Calories from Fat 5
% Daily Value	
Total Fat 0.5g	1%
Saturated Fat 0g	0%
Trans Fat 0g	
Cholesterol 0mg	0%
Sodium 0mg	0%
Total Carbohydrate 21g	7%
Dietary Fiber 4g	16%
Sugars 1g	
Protein 4g	
Vitamin A 0% * Vitamin C 0%	
Calcium 0% * Iron 2%	

Percent Daily Values (DV) are based on a 2,000 calorie diet. Your daily values may be higher or lower depending on your calorie needs.

	Calories	2,000	2,500
Total Fat	Less Than	65g	80g
Sat Fat	Less Than	20g	25g
Cholesterol	Less Than	300mg	300mg
Sodium	Less Than	2,400mg	2,400mg
Total Carbohydrate		300g	375g
Dietary Fiber		25g	30g
Protein		50g	65g

Calories per gram
Fat 9 * Carbohydrate 4 * Protein 4

Ingredients: Unbleached 100% Hard Red Whole Wheat

Nutrition Facts	
Serving size 1 egg (44g)	
Servings per container 12	
Amount per serving	
Calories 70	Calories from Fat 35
% Daily Value	
Total Fat 4g	6%
Saturated Fat 1.5g	8%
Trans Fat 0g	
Cholesterol 190mg	62%
Sodium 55mg	2%
Total Carbohydrate 1g	0%
Not a significant source of Dietary Fiber or Sugars.	
Protein 6g	10%
Vitamin A 6% * Vitamin C 0%	
Calcium 2% * Iron 4%	

Percent Daily Values (DV) are based on a 2,000 calorie diet. Your daily values may be higher or lower depending on your calorie needs.

	Calories	2,000	2,500
Total Fat	Less Than	65g	80g
Sat Fat	Less Than	20g	25g
Cholesterol	Less Than	300mg	300mg
Sodium	Less Than	2,400mg	2,400mg
Total Carbohydrate		300g	375g
Dietary Fiber		25g	30g
Protein		50g	65g

Calories per gram
Fat 9 * Carbohydrate 4 * Protein 4

Ingredients: Egg

1. There are 6 g of protein in the egg. How many calories in the egg are from protein?

2. Is whole wheat flour a good source of dietary fiber?

3. There is a type of fat not listed on the label. What is it?

4. How many more servings are there in a bag of flour than in a carton of eggs?

5. How many of the calories in a serving of whole wheat flour are from carbohydrates?

6. Which has a higher % Daily Value for protein: flour or eggs? How much higher?

Chapter 4: Activity Worksheet *continued*

Reading a Food Label

In your own words, write an explanation for each section on the food label. If you need help, you may refer to the information found at the beginning of this activity. Once completed, use this chart to teach a friend or family member how to properly read a food label.

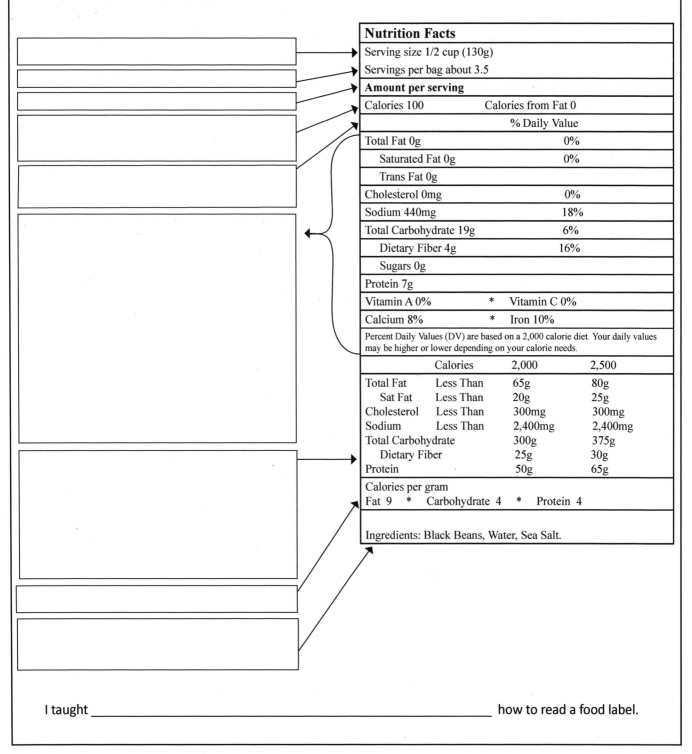

Nutrition Facts		
Serving size 1/2 cup (130g)		
Servings per bag about 3.5		
Amount per serving		
Calories 100	Calories from Fat 0	
	% Daily Value	
Total Fat 0g		0%
Saturated Fat 0g		0%
Trans Fat 0g		
Cholesterol 0mg		0%
Sodium 440mg		18%
Total Carbohydrate 19g		6%
Dietary Fiber 4g		16%
Sugars 0g		
Protein 7g		
Vitamin A 0%	*	Vitamin C 0%
Calcium 8%	*	Iron 10%

Percent Daily Values (DV) are based on a 2,000 calorie diet. Your daily values may be higher or lower depending on your calorie needs.

	Calories	2,000	2,500
Total Fat	Less Than	65g	80g
Sat Fat	Less Than	20g	25g
Cholesterol	Less Than	300mg	300mg
Sodium	Less Than	2,400mg	2,400mg
Total Carbohydrate		300g	375g
Dietary Fiber		25g	30g
Protein		50g	65g

Calories per gram
Fat 9   *   Carbohydrate 4   *   Protein 4

Ingredients: Black Beans, Water, Sea Salt.

I taught _____ how to read a food label.

Scrubbing Fiber

Chapter 4: Microscope Lab

Eating fiber-rich foods is like scrubbing your digestive system.

Have you ever used an abrasive cleaner to scrub a grimy tub? The rough particles in the cleaner scrape off the grime and take it away when you rinse. Your tub is clean.

Dietary fiber is like an abrasive cleaner. Your body cannot use some of the molecules present in the food you eat. These molecules can gunk up your insides. When you eat foods rich in dietary fiber, the dietary fiber goes through you like an abrasive cleaner. The fiber takes the "gunk" with it and leaves you clean inside.

Dietary fiber is a type of carbohydrate that comes from plants. The best sources of dietary fiber are peas, beans, whole fruits, vegetables, nuts, and whole grains. When grains are picked, they have the bran, germ, and endosperm. The bran and germ have a lot of dietary fiber in them. Refined grains, like all-purpose flour, have been sifted to remove the bran and germ, leaving only the soft endosperm and very little fiber.

Whole-grain flour is made by crunching up the bran, germ, and endosperm. Whole-wheat flour is a whole-grain flour.
Refined-grain flour is made by crunching up the bran, germ, and endosperm and then removing the bran and germ.
All-purpose and white flour are refined-grain flours.

Chapter 4: Microscope Lab *continued*

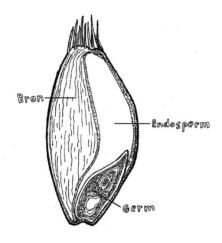

Today you are going to look at whole-wheat and all-purpose flour under the microscope. Both are made from the same starting grain: wheat. You will be able to see that the whole-wheat flour has more dietary fiber in it. Just imagine how much better it is at scrubbing the gunk out of you.

While we're at it, let's do some chemistry too! Let's do the starch test. Starch is in potatoes, corn, rice, bananas, and wheat. It is a type of carbohydrate. Whole-wheat and all-purpose flour both contain starch. Iodine is a chemical that can be used to detect the presence of starch. The starch test is where liquid iodine is added to something to test for the presence of starch. If starch is present, a chemical reaction occurs between the starch and iodine. It is easy to see because the color of the reactants is different from the color of the products. *Reactants* are the molecules at the start of a chemical reaction. They *react* with each other to form the products. *Products* are the molecules at the end of a chemical reaction. The chemical reaction between iodine and starch makes iodine a good microscope stain when looking for starch.

In this lab, you will be performing both physical and chemical tests. *Physical tests* examine the physical properties of things. Physical properties are how things look, smell, taste, and feel. *Chemical tests* examine the chemical properties of things. Chemical properties are how things behave in chemical reactions.

Materials

- Label from one bag of whole-wheat flour
- Label from one bag of white all-purpose flour
- 1 tsp whole-wheat flour
- 1 tsp white all-purpose flour
- Teaspoon measuring spoon
- Butter knife
- 4 slides
- 2 slide covers

- Syringe
- Water
- 2 toothpicks
- 2 glass bowls
- Microscope
- 4 to 6 drops of iodine
- Colored pencils

Chapter 4: Microscope Lab *continued*

Procedure

1. Read the label to determine the amount of dietary fiber in each type of flour. Record your findings on the lab sheet.

2. Measure 1 teaspoon of each type of flour into separate glass bowls. Clean off the spoon when going from one type of flour to the next. Pinch a sample of each flour type between your fingers to determine if you can feel a difference between them. Taste, look at, and smell each type of flour. These are physical tests you are performing on the two flour types. Record your observations from these physical tests on your lab sheet.

3. Use a butter knife to smear a small amount of each flour type in a thin layer on separate slides, for a total of two slides. Do not put a slide cover on top of the flour. The light from the base of the microscope needs to shine through the sample so make sure you do not have too much flour on the slide.

4. Make sure you have the nosepiece turned so the 40x lens is focusing on the stage. Carefully, so you do not spill flour from the slide onto the microscope, put the slide on the stage. Look at both slides. Record your observations on the lab sheet. The higher power objective lenses are not used with this experiment.

5. The starch test: Divide the flour in both bowls into two piles each, making four piles of flour total. Syringe up some iodine. Place two to three drops of iodine onto **one** pile in each of the bowls. The number of drops depends on the size of the drops. Do not drench the sample. Using a separate toothpick for each flour type, mix the flour and iodine samples. Take a small amount of the flour iodine mixture and smear it on the side of the bowl. Lightly rub iodine from the tip of the syringe next to, but not into, this smear. Compare the color of the iodine, the color of the unstained flour, and the color of the flour iodine mixture. Record your results and observations.

6. Clean the syringe. Place one drop of water onto one of the two unused slides. With the toothpick, take a sample for one of the flour types of the iodine flour mixture and put it into the water on the side. Mix this using the toothpick to break up any clumps. Put a slide cover on the slide. Look at this slide under the microscope at 40x. Record your results and observations. Do the same thing for the other flour type.

7. Complete the lab sheet.

Scrubbing Fiber

Chapter 4: Microscope Lab Sheet

Name_____ **Date**_____

Dietary Fiber. Record the amount of dietary fiber found in each type of flour. (This information is on the food label found on the bag.)

_____ g fiber **Whole-Wheat**   _____ g fiber **All-Purpose**

Physical Tests. Note the differences and commonalities:

Feel		
Taste		
Sight		
Smell		

Starch Test. Record the color.

	Whole-Wheat	All-Purpose
Flour		
Iodine		
Flour & Iodine		

Conclusion. Did you confirm the presence of starch in the two types of flour?

Chapter 4: Microscope Lab Sheet *continued*

When you looked through the microscope, had all parts of the whole-wheat flour changed color after the starch test?

What about the all-purpose flour?

Whole-wheat flour has the bran and germ; all-purpose flour does not. Both flour types have the endosperm in them. What do the views under the microscope, after the starch test, tell you about the starch content of the bran and germ versus the endosperm?

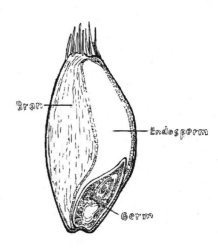

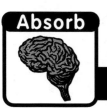

Absorb Cholesterol

Chapter 4: Famous Science Series

normal artery

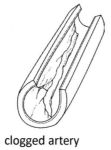

clogged artery

Cholesterol is a type of fat molecule found in the bloodstream and in all the cells of your body. Cholesterol is an important part of a healthy body, but too much cholesterol in your blood is bad for you. When people have too much cholesterol, we say their cholesterol is too high. A bloodstream too high in cholesterol can lead to clogged arteries.

Research cholesterol to answer the following questions:

What are the sources of cholesterol?

Is cholesterol found in plants?

What are LDL and HDL? Which is bad and which is good?

What can you do to avoid having high levels of bad cholesterol?

Pandia PRESS

The Chemistry of Biology

Chapter 4: Show What You Know

Questions

In 10 water molecules (10 H_2O), how many hydrogen atoms are there?

How many oxygen atoms are there?

What are the six main elements organisms are made from?

1.

2.

3.

4.

5.

6.

Multiple Choice

1. These molecules are the main source of energy for your body:

    ○ proteins
    ○ carbohydrates
    ○ nucleic acids
    ○ lipids
    ○ water
    ○ vitamins and minerals

2. Calcium makes bones and teeth strong. Calcium is a

    ○ protein.
    ○ carbohydrate.
    ○ nucleic acid.
    ○ lipid.
    ○ mineral.

Chapter 4: Show What You Know *continued*

3. Skin, hair, and hemoglobin are all made from these molecules:

   ○ proteins
   ○ carbohydrates
   ○ nucleic acids
   ○ lipids
   ○ water
   ○ vitamins and minerals

4. These molecules make DNA:

   ○ proteins
   ○ carbohydrates
   ○ nucleic acids
   ○ lipids
   ○ water
   ○ vitamins and minerals

5. These are the molecules that make proteins:

   ○ carbohydrates
   ○ amino acids
   ○ lipids
   ○ nucleic acids
   ○ vitamins and minerals

6. These molecules are used to make cell membranes:

   ○ proteins
   ○ carbohydrates
   ○ nucleic acids
   ○ lipids
   ○ water
   ○ vitamins and minerals

7. What molecules are needed by your body for chemical reactions to occur in?

   ○ proteins
   ○ carbohydrates
   ○ nucleic acids
   ○ lipids
   ○ water
   ○ vitamins and minerals

8. What are cells made from?

   ○ atoms
   ○ molecules
   ○ cytoplasm and organelles
   ○ all of the above

9. What is a group of all the same atoms is called?

   ○ an element
   ○ a bond
   ○ carbon
   ○ a molecule

10. What are the links between atoms called?

   ○ elements
   ○ molecules
   ○ bonds
   ○ nucleic acids

Chapter 4: Show What You Know *continued*

11. The amount of water in your body is

○ 50%.

○ 25%.

○ 15%.

○ 65%.

Use this food label from a box of macaroni and cheese to answer the following multiple choice questions.

12. What is the serving size of macaroni and cheese?

○ ⅓ cup

○ 270

○ ⅔ cup

○ 44g

Nutrition Facts
Serving size 2/3 cup (68g)
Servings per container: About 3
Amount per serving

Calories 270	Calories from Fat 50

	% Daily Value
Total Fat 5g	8%
Saturated Fat 3g	14%
Trans Fat 0g	
Cholesterol 15mg	5%
Sodium 480mg	20%
Total Carbohydrate 44g	15%
Dietary Fiber 2g	7%
Sugars 0g	
Protein 11g	

Vitamin A 2%	*	Vitamin C 0%	
Calcium 10%	*	Iron 10%	

Percent Daily Values (DV) are based on a 2,000 calorie diet. Your daily values may be higher or lower depending on your calorie needs.

	Calories	2,000	2,500
Total Fat	Less Than	65g	80g
Sat Fat	Less Than	20g	25g
Cholesterol	Less Than	300mg	300mg
Sodium	Less Than	2,400mg	2,400mg
Total Carbohydrate		300g	375g
Dietary Fiber		25g	30g
Protein		50g	65g

Calories per gram
Fat 9　*　Carbohydrate 4　*　Protein 4

Ingredients: Durum wheat semolina elbows, cheddar cheese (milk, salt, cheese cultures, enzymes), whey, salt, lactic acid, citric acid, natural mixed tocopherols

13. If you eat the entire box, how many calories would you consume?

○ 270

○ 540

○ 810

○ 150

14. What is the second most common ingredient in macaroni and cheese?

○ durum wheat semolina elbows

○ cheddar cheese

○ salt

○ whey

Bonus: There is cholesterol in the macaroni and cheese. What ingredient did it come from? Could this product be made cholesterol-free? If yes, explain how.

Chapter 5: Let's Get Things Moving

 **Explore** # Diffusion Confusion
Chapter 5: Lab

This lab is conducted over two days.

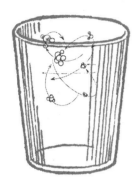

The cell membrane is made from lipid and protein molecules bonded together to enclose the inside of each cell. These molecules are a tight-knit group. They like to hang out together. There is just enough space between the molecules of the cell membrane, for small molecules like oxygen (O_2), carbon dioxide (CO_2), and water (H_2O), to diffuse between them in and out of the cell. Big molecules, like starch, $C_6H_{10}O_5$, can't diffuse across the cell membrane.

The plastic that baggies are made from is also a semi-permeable membrane. Baggies are made from molecules. The molecules that baggies are made from have spaces between them. Some small molecules can diffuse through the bag, just as they do through your cell membrane.

In this experiment, you'll make two different solutions. The first is a starch solution made with cornstarch and water, this solution is sealed in the baggie. The second solution is made from iodine and water, this solution is poured in the glass that surrounds the sealed baggie.

Cornstarch is a type of starch. When iodine and cornstarch mix, you get a positive result for the starch test. If iodine molecules diffuse through the membrane into the baggie, the cornstarch solution in the bag will change color. If cornstarch molecules diffuse through the membrane out of the baggie, the iodine solution outside the baggie will change color.

How big are the holes in baggies? Iodine molecules are small. Will iodine diffuse across the plastic membrane of the baggie? Cornstarch molecules are larger. Will cornstarch diffuse across the membrane? Let's experiment and find out. You will be using the starch test to determine the answer.

Materials

- Measuring cup
- Tablespoon measuring spoon
- Teaspoon measuring spoon
- 2 tablespoons (T) cornstarch

- 2 teaspoons (t) iodine
- Water
- 1 small, sealable plastic baggie
- 1 wide, clear glass or a measuring cup

- Notebook paper

Chapter 5: Lab *continued*

Procedure

1. On the lab report, write the hypothesis for this experiment before performing it.

2. Stir 2 T of cornstarch into 1 cup of water. Pour this into the baggie. Seal the baggie. Remove as much of the air as possible when you do this. Rinse off the outside of the baggie after it has been sealed so there is no cornstarch on the outside of the baggie. Put the baggie into the glass.

3. Fill the glass ¾ full with water. Measure 2 t of iodine into the glass. Watch as the iodine diffuses throughout the solution.

4. Write down the starting time of your experiment. This is the time when you put the iodine in the glass.

5. Check the solution for color change every ten minutes for thirty minutes. Make notes of your observations.

6. After that, check the solutions every hour three times. Make notes of your observations.

7. Let the baggie sit in the iodine overnight or longer. It takes a couple of hours before you can really tell the experiment is working. The diffusion process in this experiment is slow.

8. The next morning, check the solutions. Make notes of your observations.

9. When you are done monitoring the solution for color change, make drawings on the back of your lab report showing what was happening across the membrane of the bag. Include the movement of the molecules in your drawing.

10. When you are through, remove the baggie and stir 1 T cornstarch into the iodine solution in the glass. This is to see what would have happened if any cornstarch had diffused through the membrane of the bag into the iodine solution.

11. Complete your lab report.

Diffusion Confusion

Chapter 5: Lab Report

Name_____ **Date**_____

Hypothesis

Procedure

Observations

(see drawings)

Results and Calculations

Conclusions

Chapter 5: Lab Report *continued*

Observation Drawings

 **Explore**

In Search of Starch
Chapter 5: Microscope Lab

Cornstarch molecules are too big to diffuse across cell membranes or through the semi-permeable membrane of a plastic bag. How big are they? Could you see them with a microscope?

In this lab, you will stain a corn kernel with iodine. If there are starch molecules in the kernel, the iodine will react with them. You will make a slice of the corn kernel, but it is still thick. It can be hard to get enough light from below when you are using a thickly sliced specimen. You compensate for this by using top lighting. Top lighting refers to a light source coming from above the specimen. A desk lamp with a movable head or a flashlight with a bright, direct beam can be used.

Materials

- Microscope
- Flashlight or desk lamp
- Another person, very helpful if using a flashlight

- One kernel of fresh corn
- 1 drop iodine
- Small dish
- Tweezers with pointed ends

- X-Acto knife
- Cutting board
- Water
- Slide

Procedure

1. Using the cutting board and X-Acto knife, slice a small sliver from the kernel. This is to remove the skin so you can stain the inside.

2. Drip 1 drop of iodine into the small dish. Put the kernel of corn cut-side down in the iodine. Wait one minute. With the tweezers or your fingers, pick up the kernel and turn it over. Rinse the kernel in water to get off the excess iodine.

 Pandia PRESS

Chapter 5: Microscope Lab *continued*

3. Have there been any changes to the kernel visible with the naked eye? If so, write this in the comments section of the view sheet.

4. Put the kernel back on the cutting board. Hold it with the tweezers, <u>not</u> your fingers, as you make another slice of the corn. You want to slice the cut side that you stained with iodine as thinly as possible. BE CAREFUL!

5. Put the slice, stained-side up, on a slide. You do not need a slide cover or water. Put the slide on the microscope stage. <u>Do not let the lens touch the specimen.</u>

6. At 40x magnification, experiment with bottom and top lighting. Discover which lighting gives you the better view. You will need to play around with the placement of the light from the top lighting. It needs to be shining on the specimen.

7. Draw the view at 40x magnification. The higher magnifications are not used for this experiment. Complete the view sheet.

In Search of Starch

Chapter 5: Microscope View Sheet

Name_____ **Date**_____

Specimen _____

Type of mount_____ Type of stain used_____

**40x corn kernel slice
with starch stain**

1. Did you get a positive result with the starch test?

2. Did you see cornstarch molecules? If yes, what color were they?

3. Was the view better with top lighting?

Absorb

Secondhand Smoke

Chapter 5: Famous Science Series

What is secondhand smoking?

What is the process called that carries the smoke from its source to other areas?

What are the health risks associated with secondhand smoke?

Pandia PRESS

Let's Get Things Moving
Chapter 5: Show What You Know

1. If you put 5 drops of blue food color in a glass of water, what would happen to the food color? What is this process called? Does it require energy? In the space below, draw two pictures of the glass, one right after the food color was dripped into the glass and another an hour later.

2. In the above scenario, the food color moved from an area of _____ concentration to an area of _____ concentration.

3. Tree roots use passive transport to absorb the water the tree needs. What is this type of passive transport called? _____

4. During active transport, _____ is needed to move molecules from an area of _____ concentration to an area of _____ concentration.

5. Two examples of active transport are _____ and _____.

6. If your mom cleans your room, do you use your energy to do it? Is it active for you, or passive?

7. What if you clean your room? Is this active or passive?

Chapter 5: Show What You Know *continued*

8. Choose the best word from the word list that defines each statement.

> | | | | | |
> |---|---|---|---|---|
> | Endocytosis | Exocytosis | Hydrophilic | Hydrophobic | Semi-permeable membrane |
> | Lipid bilayer | Osmosis | Active transport | Passive transport | Diffusion |

_____ This is the process where cells use energy to move molecules from an area of low concentration to an area of high concentration.

_____ This is the process where molecules move from an area of high concentration to an area of low concentration.

_____ A process that transports large molecules into the cell. Material comes in contact with the cell. Then it is enclosed within its own membrane. Next, it is brought inside the cell and released from the membrane into the cell.

_____ A process that transports large molecules out of the cell. Material inside the cell is enclosed within its own membrane. It is taken to the cell membrane, with which it then fuses. After that, the material is released out of the cell.

_____ Water-loving

_____ Water-hating

_____ Made of two layers of lipids.

_____ The diffusion of water across a membrane.

_____ The process of moving molecules into and out of cells without using any cellular energy. Diffusion is an example of this.

_____ Lets some but not all things pass through it.

9. Which three of the nine characteristics of life directly relate to the cellular transport? Give a one-sentence explanation for how each works.

   1.

   2.

   3.

Chapter 6: Cell Energy

Energy In, Energy Out

Chapter 6: Lab

The best place to put the energy cycle that fuels life on planet Earth in perspective is outside on a sunny day. For today's lab, you are going to go outside and do just that.

You cannot see sunlight, but you can feel it. Some chemical reactions need energy. The chemical reaction that starts photosynthesis is one of them. Plants get the energy they need to make their own food from the sun. Just imagine it. Wow! Making your own food without eating, just standing in the sunshine.

When you are outside you will look at both the *macro-scale processes* and the *micro-scale processes*. Macro-scale processes look at things that happen on a large (macro) scale, such as the whole cycle of molecules and energy in photosynthesis. Micro-scale processes look at things that happen on a very small (micro) scale, such as the chemical reactions involved in photosynthesis.

Materials

- Pencils in various colors, make sure one is green
- Sunny day, preferable
- Fruit or vegetable snack, to eat when instructed
- Glue and construction paper (optional)

Procedure

1. Go outside and feel the sunshine on you. Feel the warmth and energy from the sun. Now look at some plants. Imagine what is happening to those plants. It looks like plants are just there, doesn't it? It might not look like plants are busy going through a series of chemical processes that are the starting point for meeting all the energy needs of the entire planet, but they are.

Chapter 6: Lab *continued*

2. Take out the Micro-Scale Lab Sheet.

- Think about the sunlight shining on chloroplasts in the green part of the plant.

- Chloroplasts absorb the energy from the sun and turn it into energy cells can use.

- There is carbon dioxide entering the leaves diffusing in through ***stomata***. Stomata* are small openings or pores in leaves that let small molecules, such as oxygen and carbon dioxide, diffuse into and out of the leaf. *The singular form of stomata is stoma.

- Water is being transported from the roots to the leaves.

3. Find a plant with a leaf and look at the leaf. As you look at the leaf, it is working away making glucose and oxygen. Use the leaf drawing on the Micro-Scale Lab Sheet, or turn the lab sheet over and draw a picture of the leaf you are observing. On the lab sheet:

- Make a diagram on the leaf of the micro-scale processes that are happening. You will be diagramming the process of photosynthesis.

- With the green pencil, color in the ovals on the leaf. These are the chloroplasts.

- If you drew your own leaf, draw one or more stomata on the leaf.

- Color the circle in the top right of your lab sheet yellow. This is the sun.

- Choose one chloroplast, draw sunlight shining on that chloroplast. Label the arrow "sunlight = energy."

- Draw all the molecules going to or leaving from this chloroplast:

  ➤ Draw six water molecules traveling toward the chloroplast from the stem. Use arrows to show where the water molecules are going. You can use the symbol for water, H_2O, or blue dots.

  ➤ Draw six carbon dioxide molecules coming through the stomata on the leaf going to the chloroplast. Draw each CO_2 molecule outside of the leaf. Draw an arrow from each molecule to a stoma. You can use a different stoma for each molecule or you can draw multiple CO_2 molecules using the same stoma. Draw another arrow from the stoma to the chloroplast. You can use the symbol for carbon dioxide, CO_2, or a different-colored dot to represent the molecules.

  ➤ Draw a glucose molecule with an arrow coming from the chloroplast. The glucose molecule stays in the plant; do not draw it leaving the plant. You can use the symbol for glucose, $C_6H_{12}O_6$, or a different-colored dot to represent the molecule. After you make the symbol for glucose, write "= energy" so that at the end of the arrow it says "$C_6H_{12}O_6$ = energy."

  ➤ Draw six arrows leaving the chloroplast going to one or more stomata. Draw arrows going from the stomata to the outside of the leaf. At the point of each arrow outside of the leaf, draw an oxygen molecule. You can use the symbol for oxygen, O_2, or different-colored dots to represent the molecules.

- Fill in the key on the lab sheet. Identify what symbols, whether chemical or colored dots, you used for each chemical in your representation of the process of photosynthesis.

Chapter 6: Lab *continued*

4. Take out the Macro-Scale Lab Sheet. Think about the sun shining on the plants and you. Think about the energy from the sunlight and how plants turn this chemical energy into usable energy for all life on the planet.

5. Take the leaf that is still attached to a plant in your hand. Put your mouth right next to the leaf. Breathe in and breathe out on the leaf. Breathe in and think about the fact that there are oxygen molecules in your lungs right now that just came from the plants around you. Breathe out. There were carbon dioxide molecules in the breath you exhaled that will be inside the plants around you, being turned into glucose molecules. Some of the glucose molecules made by the plant will be eaten by other organisms. If you have a garden, these plants might even be eaten by you. Animals will use these glucose molecules for energy. When the glucose molecules are used for energy, one of the products is CO_2. Animals breathe out CO_2 and the cycle keeps going. You can use the drawing on the lab sheet, or turn the lab sheet over and draw a picture of what you are observing. Leave room for labels and arrows. On the lab sheet:

- Draw yourself in the picture.

- Draw animals in the picture.

- Draw an arrow from the sun to the tree and plants. Label the arrow "energy."

- Draw arrows from the tree and plants to you and the animals. Label the arrows "oxygen" or "O_2."

- Draw arrows from you and the animals to the tree and plants. Label the arrows "carbon dioxide" or "CO_2."

- Draw an arrow going from the ground up the trunk to the leaves of the tree. Label the arrow "water" or "H_2O."

- At the top of the tree among the leaves, write "plants make glucose."

- Remember, insects and worms are animals; if you didn't already, draw some on the lab sheet eating plants, including tree leaves.

- Eat the fruit or vegetables you brought with you. Draw this on the lab sheet. Draw an arrow from the plant food you are eating into your stomach. Label this arrow "glucose." For any animals that are large enough, draw arrows from the plants the animals are eating in toward their stomachs. Label these arrows "glucose."

Energy In, Energy Out

Chapter 6: Micro-Scale Lab Sheet

Name_____ **Date**_____

water =

carbon dioxide =

glucose =

oxygen =

stoma =

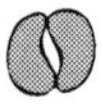

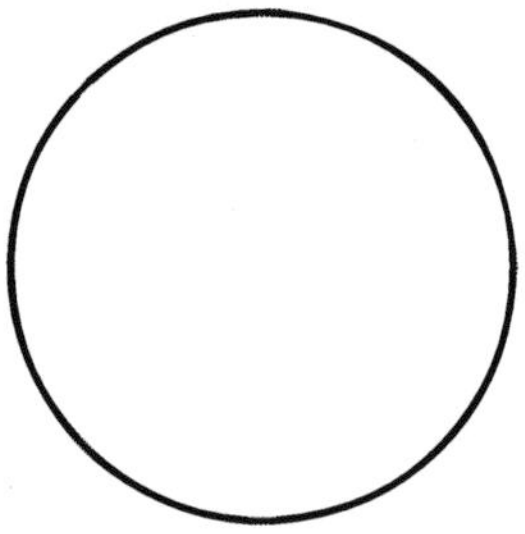

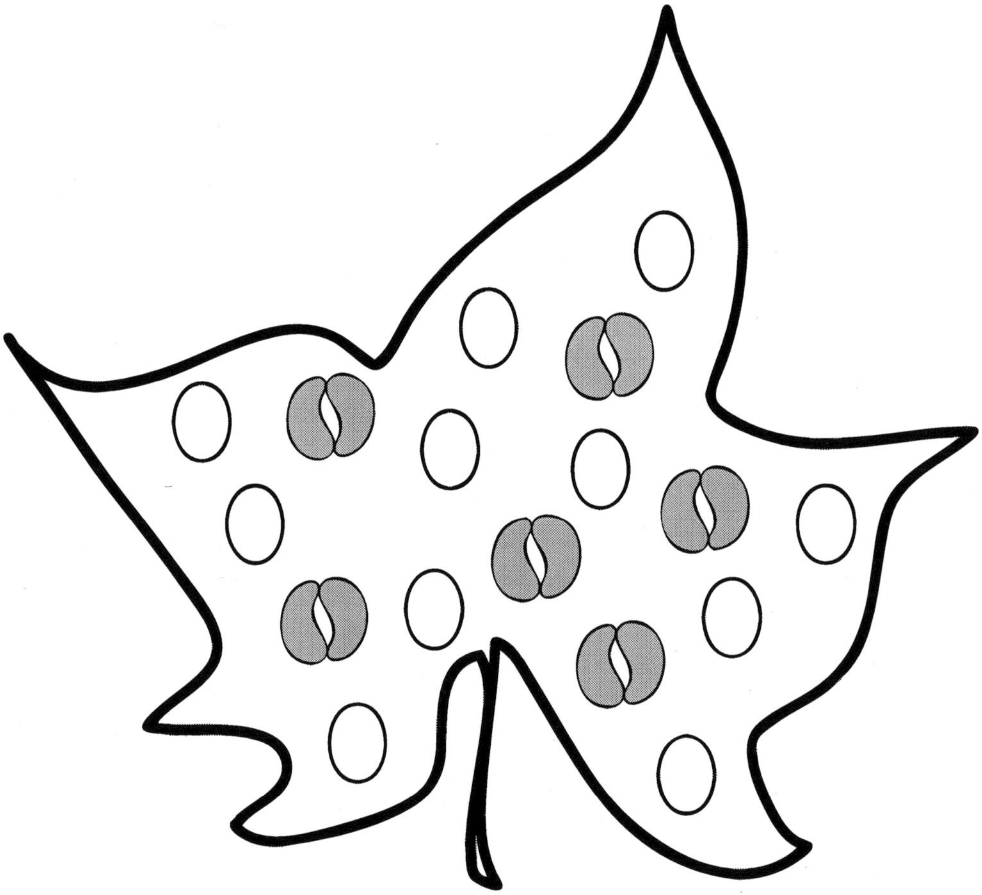

Pandia PRESS

Energy In, Energy Out

Chapter 6: Macro-Scale Lab Sheet

Name_____ Date_____

Going Green

Chapter 6: Microscope Lab

You hear the term "going green" a lot these days. Plants don't have to go green; they already are. Chloroplasts are organelles in the cells of plants that absorb the sunlight. They convert sunlight into energy they can use. When you look at a green leaf, you are looking at lots of chloroplasts. If you look at a green leaf through a microscope, you can see the individual chloroplasts. You can look at the original solar panels.

Materials

- One leaf from a thick and/or stiff-leafed plant, such as a succulent or an orchid
- X-Acto knife
- Microscope
- Slide
- Slide cover
- Water

- Syringe
- Green and gray pencil
- Oil, for oil immersion, binocular microscope only
- Cleaner for oil
- Microscope View Sheet (choose the one that matches your scope)

Procedure

1. Cut a thin slice of the leaf along the top side. You need a small square about ½ cm by ½ cm. You might need to make several slices until you get a thin enough slice. If the slice is too thick, cells are stacked on top of each other, which leads to an unclear view of them.

2. Put it on the slide. Drip one to two drops of water on the slide.

Chapter 6: Microscope Lab *continued*

3. Put on the slide cover.

4. Start at 40x, find an area where the specimen is thin and focus on it.

5. Turn the nosepiece so the 100x objective lens is focused on the specimen. Find a group of chloroplasts that looks interesting.

6. Turn to the 400x objective lens. Draw this view on your view sheet. Use green for the chloroplasts and gray for the rest. Draw the cell walls of all the cells, but only draw the chloroplasts for one cell.

7. Binocular microscope only: Lower the stage as low as it will go. CAREFULLY drip one drop of the oil onto the slide cover at about the place where the lens is focused. Do not move the slide! Do not move the stage! Raise the stage and focus the 1000x (oil immersion) objective lens. Do not use the stage knobs when you raise the stage. Draw the view as seen through the 1000x objective lens. You may need to go back to the 100x lens and focus with the pointer on a group of chloroplasts, and then come back to the 1000x lens. Do not use the 400x lens to refocus. You might get oil on the lens and that could ruin it. Use the view sheet included with this lab to draw one chloroplast, use green. When you are done, wipe the oil immersion lens with a lens wipe and cleaner.

Going Green

Chapter 6: Microscope View Sheet

Name_____ **Date**_____

Monocular Microscope

Specimen _____

Type of mount_____ Type of stain used_____

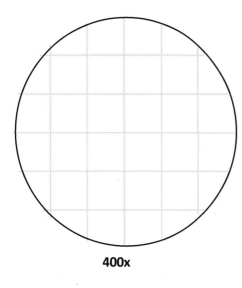

400x

1. Where are chloroplasts in the cell? Why do you think they are there?

2. Why didn't you use stain to look at the chloroplasts?

3. What is the name of the molecule that makes chloroplasts green? When you look at chloroplasts, do you think you are looking at one or more than one of these molecules? Why?

Comments:

Going Green

Chapter 6: Microscope View Sheet

Name_____ Date_____

Binocular Microscope

Specimen _____

Type of mount_____ Type of stain used_____

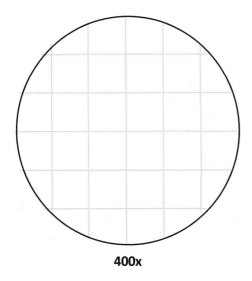
400x

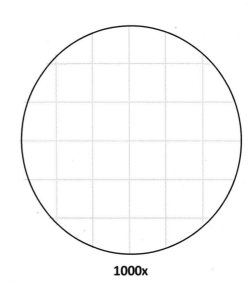
1000x

1. Where are chloroplasts in the cell? Why do you think they are there?

2. Why didn't you use stain to look at the chloroplasts?

3. What is the name of the molecule that makes chloroplasts green? When you look at chloroplasts, do you think you are looking at one or more than one of these molecules? Why?

Comments:

Absorb

Stromatolites

Chapter 6: Famous Science Series

Famous Fossilized Photosynthesizer

When you think of photosynthesis, you probably think of trees. If so, it might surprise you to learn that most photosynthesis on the planet happens in the ocean. In fact, the ocean is where photosynthesis began. Stromatolites are the oldest known fossils. Some of them date back 3.5 billion years. Stromatolites are created by huge groupings of photosynthesizing cyanobacteria, also called blue-green algae.

Draw a picture of this famous fossil. Write two or more facts about this fossil.

Facts:

 **Cell Energy**

Chapter 6: Show What You Know

Photosynthesis 411

1. Not all organisms can photosynthesize. List the group of organisms that can and state whether they are heterotrophs, autotrophs, or both.

2. In the boxes below, write the overall chemical reaction for photosynthesis.

Reactants	Products

→

3. Name the organelle where photosynthesis happens.

4. Why is this chemical reaction important?

Cellular Respiration 411

5. What organisms perform cellular respiration? Are they heterotrophs, autotrophs, or both?

6. In the boxes below, write the overall chemical reaction for cellular respiration.

Reactants	Products

→

7. Name the organelle where cellular respiration happens.

8. Why is this chemical reaction important?

Chapter 6: Show What You Know *continued*

9. Photosynthesis and cellular respiration are a called a cycle. Why?

10. Match the word with the correct definition.

Chlorophyll ○

○ The energy that is stored between the bonds in a molecule.

Fermentation ○

○ Waste molecule made during cellular respiration.

Glucose ○

○ $C_6H_{12}O_6$, the sugar molecule that is made during photosynthesis.

Cellular respiration ○

○ A type of respiration that takes place in the absence of oxygen.

Chemical energy ○

○ The main process cells use to get energy.

Carbon dioxide ○

○ Waste molecule made during photosynthesis.

Oxygen ○

○ Green molecule in plants that is able to capture energy from the sun.

11. Compare and contrast lactic acid fermentation with alcoholic fermentation.

12. What is the significance of van Helmont's experiment?

Pandia PRESS

Chapter 7: The Message

Read

What in the World Is DNA?

Chapter 7: Lesson Activity

The following coloring activity is found in Chapter 7 Lesson in your Textbook.

A Molecule of DNA

1. DNA is made from nucleotides called bases. There are four different bases in DNA molecules. *Color the bases below.*

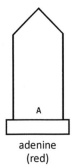
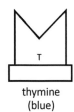
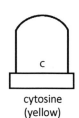
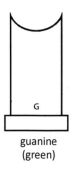

| adenine (red) | thymine (blue) | cytosine (yellow) | guanine (green) |

2. The bases pair together. In DNA, A always pairs with T, and C always pairs with G. The base pairs are held together with chemical bonds, called **hydrogen bonds**. One hydrogen bond is not very strong, but thousands of them are. It is like a zipper. *Color the base pairs below.*

adenine, A (red)

thymine, T (blue)

cytosine, C (yellow)

guanine, G (green)

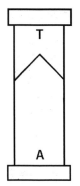
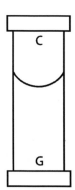

PandiaPRESS

3. A DNA molecule is long and thin. The bases in each strand are connected by sugar (S) and phosphate (P) molecules. *Color the sugar (S) purple and phosphate (P) orange.*

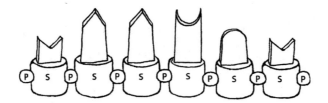

4. Each strand bonds together at the centers of the base pairs. The two strands spiral around, forming a double helix similar to a spiral staircase. The two strands together make one long molecule. The sugar and the phosphate are shown as a segmented rope on the outside of the helix.

   Color adenine (A) red, thymine (T) blue, cytosine (C) yellow, and guanine (G) green. Color the outer segmented rope alternating purple and orange to show the alternating sugars and phosphates that hold the DNA bases in place.

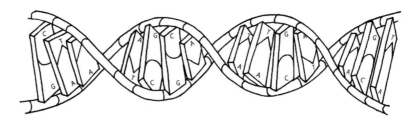

DNA Can Replicate

The base pairing of DNA gives it a special property. Using each strand as a template DNA can replicate. *Replicate* means "to copy," and that is what DNA does. It unzips, opens up, and matches its base pair along the strand. Each side makes a copy of the other side.

Replicate the DNA strand shown below, by drawing its base pair on top of each base. After you are done drawing in the new base pairs, color adenine (A) red, thymine (T) blue, cytosine (C) yellow, and guanine (G) green.

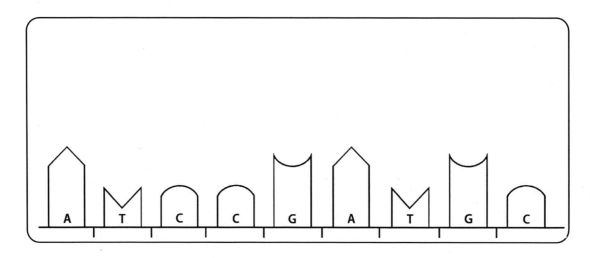

Explore

Marshmallow DNA

Chapter 7: Lab

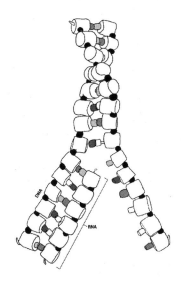

DNA molecules are double helixes, shaped like spiral staircases. When RNA molecules form along a gene in a DNA molecule or when the DNA replicates itself, the DNA uncoils and opens, so that the new molecule can form along the DNA strand. You are going to make a DNA molecule that has uncoiled partway while transcription occurs. You will build an RNA molecule along the length of DNA that is uncoiled.

DNA is a nucleic acid. It is a large molecule made from other smaller molecules. There are three different types of molecules that make DNA. The two strands that coil around the outside of a DNA molecule are made from a type of sugar molecule and a phosphate molecule. This is called the sugar phosphate backbone. These two types of molecules alternate one after the other down the DNA strands. The sugars are joined to the molecules called bases. Four different nitrogenous bases connect the two DNA strands. They pair up as adenine and thymine, and guanine and cytosine.

The single-stranded RNA molecule has a similar chemical make-up. It has a backbone made from alternating sugar and phosphate molecules. Each sugar molecule is connected to a nitrogenous base. There are four different nitrogenous bases. RNA molecules are built along the DNA strand in the following pattern:

- adenine is built at the sight on the DNA where a thymine is,
- uracil is built at the sight on the DNA where an adenine is,
- cytosine is built at the sight on the DNA where a guanine is, and
- guanine is built at the sight on the DNA where a cytosine is.

RNA molecules do not have thymine bases; they have uracil bases instead.

Materials

- Bag of multicolored mini marshmallows
- 31 beads (one color, about the size of a mini marshmallow)
- 5 twelve-inch-long pipe cleaners
- 34 large marshmallows (one color)
- Skewer
- 19 toothpicks
- Scissors

Chapter 7: Lab *continued*

Procedure

1. Choose a mini marshmallow color for each of the five bases. Below is a suggested color scheme, you may use this scheme or create your own. The amount of marshmallows needed will vary depending on how many of each base pairs you choose to use in the DNA strands you make (there are extras).

   - 12 yellow mini marshmallows = cytosine
   - 12 green mini marshmallows = guanine
   - 12 orange mini marshmallows = adenine
   - 12 white mini marshmallows = thymine
   - 4 pink mini marshmallows = uracil

large marshmallow = sugar

bead = phosphate

2. Each pipe cleaner should fit eight large marshmallows and eight beads. Begin by stringing eight large marshmallows (through the shorter side) and eight beads onto each of two pipe cleaners, alternating marshmallows and beads. Start with a marshmallow and end with a bead.

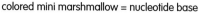

colored mini marshmallow = nucleotide base

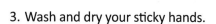

3. Wash and dry your sticky hands.

4. The large marshmallows and beads are the sugar-phosphate backbone of DNA. The large marshmallows are the sugar molecules and the beads are the phosphate molecules. At the ends of the pipe cleaners (the one with a marshmallow), there should be a little over ½ inch of pipe cleaner sticking out; fold over the pipe cleaners at these ends. The other ends should each have a bead. Set them down gently so the beads stay on the pipe cleaner.

5. Make eight nucleic acid base pairs. Do this by taking one toothpick and putting two mini marshmallows on the toothpick, pushing them until they touch in the center. Remember the bases are specific about which they pair with. Only use cytosine, guanine, adenine, and thymine at this point. Make four pairs of adenine with thymine, and four pairs of cytosine with guanine. You decide the order of the bases as you attach them.

6. Going down one of the large marshmallow chains, poke one end of each toothpick into each large marshmallow as shown.

Pandia PRESS

Chapter 7: Lab *continued*

7. Do the same thing with the other side of each toothpick and the other chain. The two chains are now one large chain with base pairs in the center. Make sure the folded-over pipe cleaner ends are both at the same end. The order the base pairs are in does not matter.

8. Twist this strand into a coil, being careful you do not pull out any toothpicks. Let it sit 15 to 30 minutes. At first, it will want to uncoil, but just keep gently recoiling it. When the marshmallows begin to dry out, it will stay coiled. You might want to coil the chain and then place something light, like a CD case, at both ends while you go on to the next part.

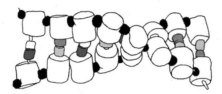

Now you are going to make a second DNA chain. Eventually, you will be attaching this second chain to the first chain you already made. In the second chain, you will build a DNA molecule that is uncoiled while transcription occurs.

9. This time, string eight large marshmallows but only seven beads onto two pipe cleaners. Alternate the marshmallows and beads, beginning with a marshmallow and ending with a marshmallow. Twist the ends of each of these pipe cleaners to the two unbent ends of the coiled pipe cleaners that you made in the steps above at the bead end of this pipe cleaner. Fold the ends of the pipe cleaner at the two ends of the pieces you just made to try to hide the twists.

10. Make one base pair using toothpicks. Do not use uracil. Poke this into the DNA strands nearest the coiled strand, connecting both sides.

11. Cut the skewer 3 inches long. Make a second base pair using the skewer. Do not use uracil. Poke this into the DNA strands, connecting both sides. There are now six marshmallows on each strand that have not been connected.

12. Using another pipe cleaner, make the RNA strand: String six marshmallows and five beads, starting and ending with a marshmallow. Fold over the pipe cleaner ends.

13. Make six base pairs for the DNA/RNA synthesis, using mini marshmallows and toothpicks, as follows (or create your own color scheme):
    • thymine–adenine, white–orange
    • adenine–uracil, orange–pink
    • guanine–cytosine, green–yellow
    • adenine–uracil, orange–pink
    • cytosine–guanine, yellow–green
    • thymine–adenine, white–orange

Chapter 7: Lab *continued*

14. Attach these bases to one of the uncoiled DNA strands in the order listed above. For example, the first base you attach should be thymine–adenine, second is adenine–uracil, etc.

15. Attach the other base-pair side to the unattached strand, the RNA sugar-phosphate backbone. The RNA strand will sit in the middle of the two uncoiled and "unzipped" DNA strands.

16. Now you need to add the bases to the other side of the DNA strand. First, cut three toothpicks in half. Top each toothpick with a mini marshmallow in the following order if you used the suggested color scheme (you do not need to poke the toothpick through the tops of these marshmallows):

   - adenine = orange
   - thymine = white
   - cytosine = yellow
   - thymine = white
   - guanine = green
   - adenine = orange

17. Carefully lay it out and let your DNA/RNA molecule model dry.

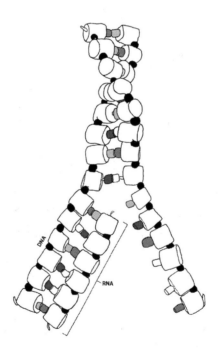

Explore Looking at My DNA
Chapter 7: Microscope Lab

★ **This lab requires some preparation the night before.**

DNA is a macromolecule, which means it is a large molecule. If you could uncoil and separate the DNA in just one of your cells, it would measure about 6 feet long. All the DNA uncoiled in your body would reach the moon!

Today you are going to look at your DNA. First, you have to collect some cells. You will do this by collecting your own cheek cells. Next, you have to break open the cell and nuclear membranes to release the DNA. You do this with dish soap. Dish detergent is very good at breaking apart the bonds holding the lipid molecules together in your cell and nuclear membranes. The DNA in your cells is wrapped around protein molecules. To get nice long strands of DNA to look at, you need to separate the protein molecules from the DNA molecules. You will use meat tenderizer to do this. Meat tenderizer is made of enzymes that break up protein molecules. When these protein molecules break up, they separate from the DNA molecules. The next step is to add cold rubbing alcohol to the solution. The rubbing alcohol mixes with the solution then separates it, forming a top layer of liquid. The DNA molecules precipitate out of the soap solution into the rubbing alcohol. The protein and lipid molecules remain in the soap solution. You will be able to see the DNA floating in the rubbing alcohol. After that, you can collect the DNA, stain it, and look at it with your microscope. When you look at DNA under the microscope, you are looking at a macromolecule. It looks sort of like a blob. This might not be the coolest-looking specimen you have seen with your microscope, but how many people can say they have actually seen their own DNA?

Materials

- Sports drink, i.e., Gatorade
- Timer
- Rubbing alcohol
- Meat tenderizer, like Accent (just a pinch)
- Toothpick
- Dish soap (just a drop)
- Small cup or glass
- Small 15 ml test tube
- Pipette or syringe
- Slide with slide cover
- Methylene blue stain
- Water
- Microscope

Chapter 7: Microscope Lab *continued*

Procedure

1. The night before the lab, place the rubbing alcohol in the freezer to get it very cold (don't worry, unless the temperature in your freezer is below -179° F, it will not freeze).

2. Lightly swish out your mouth with water. This removes food debris from your mouth.

3. Take a mouthful of sports drink and swish vigorously for at least one minute; use your timer for this. You need to collect many cheek cells so that you have many DNA molecules. This takes a minute or more of vigorous swishing.

4. Spit the mouthful of sports drink into the cup.

5. Put the detergent into the test tube. Carefully pour the contents of the sports drink you spat out (ooo!) into the test tube until the test tube is half-full.

6. Add a pinch of meat tenderizer. Put your finger or a stopper over the test tube and carefully turn the test tube over back and forth ten times. BE GENTLE: Do not make bubbles with the soap or you will not be able to see the DNA.

7. Wait 15 minutes for the soap and meat tenderizer to do their job.

8. Take the alcohol out of the freezer. Carefully, pipette or syringe about 1 teaspoon of alcohol into the test tube. You do this by putting the mouth of the pipette, or syringe, against the inside wall of the test tube and gently letting the liquid drizzle in. Do not fill the test tube all the way up. Leave some headspace. Do not mix this solution. Let it sit quietly for 15 minutes.

9. Rinse out the pipette or syringe with fresh water.

10. You will see the DNA molecules at the point where the soap solution and the rubbing alcohol layer meet. Take the toothpick and wind the DNA molecules around it.

11. Smear the DNA molecule from the toothpick onto the slide. Put a drop of methylene blue on the slide. With the water from the pipette or syringe, CAREFULLY flush the excess methylene blue from the slide. Wipe the bottom of the slide (the side WITHOUT the DNA molecule on it) if it got wet. Put a drop of water on the slide, if you need one. Cover the DNA molecule with the slide cover. Put the slide cover on the slide over your DNA molecule. Now look at your own DNA. Look at it at all magnifications, except with the oil immersion lens.

12. Choose your favorite magnification for your DNA and draw what you see on the lab sheet.

Looking at My DNA

Chapter 7: Microscope View Sheet

Name_____ **Date**_____

Specimen_____ Type of Microscope _____

Type of Stain_____ Magnification _____

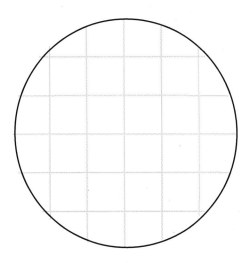

Comments:

Watson and Crick
Chapter 7: Famous Science Series

James D. Watson, Francis Crick, and Maurice Wilkins won the Nobel Prize in 1962. What did they do to earn this award?

Why wasn't Rosalind Franklin part of the group who won the Nobel Prize? Do you think that is fair?

When and where did Watson and Crick meet?

Chapter 7: Famous Science Series *continued*

James D. Watson

When and where was he born?

He had a special kind of memory. What was it?

He was a regular contestant on what radio show? When was his first appearance?

How old was he when he started college? What college did he attend?

Francis Crick

When and where was he born?

How old was he when he told his mother he wanted to be a scientist?

What was The Blitz and how did it affect Crick's work?

The Message

Chapter 7: Show What You Know

Multiple Choice

1. Base pairs in DNA connect to each other with

    ○ molecules.
    ○ chemical reactions.
    ○ hydrogen bonds.
    ○ zippers.

2. Bases in DNA are held together with a backbone of

    ○ protein chains.
    ○ sugar phosphate chains.
    ○ sulfur phosphorus chain.
    ○ RNA molecule.

3. A chromosome is made of

    ○ one molecule of DNA.
    ○ many molecules of DNA, each one gene long.
    ○ DNA and RNA molecules.
    ○ two molecules of DNA.

4. DNA is shaped like

    ○ an X.
    ○ a rope.
    ○ a string.
    ○ a double helix.

5. One gene has instructions for making

    ○ one protein.
    ○ one amino acid.
    ○ one cell.
    ○ one codon.

6. One RNA molecule codes for the synthesis of

    ○ one protein.
    ○ one amino acid.
    ○ one cell.
    ○ one codon.

7. One codon has the instructions for making

    ○ one protein.
    ○ one amino acid.
    ○ one cell.
    ○ one base pair.

8. The DNA in your cells

    ○ depends on the cells' specialization.
    ○ is the same as your parents.
    ○ is different than your parents.
    ○ is the same for all humans.

Chapter 7: Show What You Know *continued*

9. To make a protein

   ○ the chromosome unravels.

   ○ one gene that codes for making the protein unravels.

   ○ the strand of DNA unravels.

   ○ DNA leaves the cell to translate the information to the ribosome.

10. The transcription of a molecule of RNA is most like the process of

    ○ translation.

    ○ protein synthesis.

    ○ hydrogen bonding.

    ○ replication.

11. Circle the organelles that are involved in protein synthesis.

    mitochondria   ribosomes   chloroplasts

    vacuoles   Golgi apparatus

    smooth endoplasmic reticulum   rough endoplasmic reticulum

12. One hemoglobin molecule has 574 amino acids in it. How many base pairs does it take to make the amino acids in a hemoglobin molecule? How many codons?

13. The opposite strand of DNA is called the complementary strand. Make the complementary strand of DNA below.

    A C G T T A G C C G A T

14. Make a complementary strand of RNA.

    A C G T T A G C C G A T

15. Put these in order as you build a chromosome.

    gene, nucleotide base, codon, base pair ⟶ chromosome

Chapter 7: Show What You Know *continued*

16. Write definitions for the following terms:

    Replication:

    Transcription:

    Translation:

17. List four reasons RNA is used for protein synthesis.

    1.

    2.

    3.

    4.

Bonus Questions

Do the ribosomes in humans make all 20 amino acids needed to make all the 100,000 types of proteins your body makes? Support your answer.

Plants also use 20 different amino acids to build proteins. How do plants get all the amino acids they need?

Pandia PRESS

Chapter 8: Mitosis–One Makes Two

Explore

Life Cycle of a Cell Poster
Chapter 8: Lab

Today you are going to make a poster showing the life cycle of a cell.

Some things to be careful of:

- If your hands get sticky, wash them right away. You might have to do this several times while making the poster.
- Do not be stingy with the glue.
- If using cereal, be careful if you set the poster face down. The cereal will break apart.
- Make sure you glue all chromosomes INSIDE the nuclear membrane if there is one, or the cell membrane if there is not.

Materials

- 2 pieces of 14" x 22" poster board
- 4 pipe cleaners, of two different colors that are not the same color as the poster board
- 6 mini marshmallows, if using a white poster board choose marshmallows that are not white

- Markers
- Ruler
- Computer and printer (optional for typing the labels)
- CD
- Glue
- Can or jar that is 3 inches around in diameter

- Something with a 1½- to 2-inch diameter circular bottom like the bottom of a glass or a pudding or fruit cup
- Brightly colored yarn
- 6 small beads (6 mm x 9 mm), pieces of cereal will work too (e.g., Cheerios)
- Mitosis illustration from the textbook lesson

Procedure

1. On one poster board, measure down 1½ inches from the top and lightly mark this with a pencil.

2. Above this mark, write on the poster (or you can type, print, cut out, and glue) the title: The Life Cycle of a Cell. Refer to the example posters found at the end of this lab.

3. Measure down 1½ inches from that mark and lightly make another mark with a pencil.

Chapter 8: Lab *continued*

4. Above this mark, on the left side of the poster, write or make a label on your computer that says *Interphase.*

5. Below this title, with a marker, trace a circle around the outside of a CD. Inside this circle, trace around the bottom of the can.

6. Measure and cut six pieces of pipe cleaner, three of each color, 3 inches long.

7. Fold one piece of each color pipe cleaner at approximately every ¼ inch spot, back and forth. Glue these two pieces inside the smaller of the two circles you drew.

8. Draw or make a label of a short arrow pointing away from the circle.

9. Write or make a label that says *Mitosis.*

10. Draw or make a label of a short arrow pointing away from the word *Mitosis.*

11. Write or make a label that says *Cytokinesis.* Do this at the same distance from the top that the heading *Interphase* is. Make sure the heading *Cytokinesis* and the circles you draw next are not all the way over to the right side.

12. With the can, make two circles. Make a circle inside of each of these circles using the pudding cup. Fold the four pieces of pipe cleaner back and forth, as you did before. Glue one of each color, for a total of two in each of the smaller circles.

13. Draw or make a label of a short arrow pointing away from the two circles.

14. Write the word INTERPHASE along the right side of the poster, like this

I

N

T

E

R

P

H

A

S

E

Pandia PRESS

Chapter 8: Lab *continued*

15. Cut away any part of the poster board you did not use.

16. Begin poster 2. Along the long left side of the second poster board measure 2 inches from the top of the poster, making a light mark with the pencil at that point. Measure down 5 inches from that mark and mark this spot. Repeat this two more times. Do the same thing along the right side.

17. Using the ruler and a marker, make a line from one mark to the next so that you draw four lines across the board.

18. Make a title for the poster. You can use your computer and a printer, or you can do it by hand. The title should read *Mitosis = Nuclear Division*; this should be about 1 inch tall. In smaller letters, about ½ inch tall, write or type and print *PMAT*.

19. In the second space down, trace around the outside of the CD. Do this on the right side of the poster. Inside this circle, draw a dotted circle around the bottom of the can.

20. Measure and cut 6 pieces of pipe cleaner in each color, all 1¼ inches long.

21. Put two strands of the same color pipe cleaner through the center of each marshmallow. These are the chromosomes. Slightly bend each strand in a V so that the chromosome has an X shape.

22. The beads are **centrioles**. They are only found in animal cells. Centrioles help organize cell division in animal cells. Glue one of each color chromosome (for a total of two chromosomes) inside of the smaller circle you made with the can. Glue two beads on opposite sides of the smaller circle but inside the larger circle.

23. Write on the poster to the left of this first circle (or you can type, print, cut out, and glue):

    Prophase

    1. The DNA condenses into chromosomes.

    2. The nuclear membrane breaks down.

24. The yarn pieces are **spindle fibers**. Spindle fibers are protein structures that attach to the chromatids and pull them apart. In the third space down, trace around the outside of the CD. Do this on the right side of the poster below the first circle. Measure the distance between the beads in the circle above. Cut two pieces of yarn 1½ inches longer than this distance. Put two dots of glue in the same spot as the beads are in the circle above. Use more glue to glue the pieces of yarn in the circle. The yarn should span from one dot of glue to the other, with the ends in the glue. One piece of yarn should curve up and the other should curve down. Put a bead on each dot of glue. Glue one of each color chromosome (for a total of two chromosomes) down the center of the circle. Make sure each centromere is glued on a piece of yarn.

Chapter 8: Lab *continued*

25. Write on the poster to the left of this second circle (or you can type, print, cut out, and glue):

    Metaphase

    1. The chromosomes line up in the middle of the cell.

26. In the fourth space down, trace lightly around the outside of the CD. Make slight indents on the circle as shown on the diagram. Do this on the right side of the poster below the other circles. Measure and cut four pieces of yarn that are 1¾ inches long. Put two dots of glue in the same spot as the beads are in the circle above. Glue each strand of yarn from one of the dots of glue toward the center. It should look as if the yarn from the circle above has broken in two and pulled back, but that the pieces of yarn are in the same place as above. Put a bead on each dot of glue.

27. Take one of each color chromosome, pinch the marshmallow in the center, and pull the pipe cleaner pieces apart. Did that hurt? You just ripped the centromere apart. Make sure you grab the separate pieces of pipe cleaner. Squeeze the marshmallow half around the pipe cleaner so that it surrounds it. Glue one of each color chromosome on the end of the yarn directly across from the same color. Make sure each centromere is glued at the end of the yarn.

28. Write on the poster to the left of this third circle (or you can type, print, cut out, and glue):

    Anaphase

    1. The chromosomes split in two at the centromere.

    2. The separated chromosomes are pulled to opposite sides of the cell.

29. In the bottom space, trace lightly around the outside of the CD. Make slightly bigger indents on the circle as shown on the diagram below. Do this on the right side of the poster below the other circles. Inside the circle, on each side of the indent, trace a 1½- to 2-inch circle, using the pudding cup bottom as a guide.

30. Cut four 3-inch-long pieces of pipe cleaner, two in each color. Fold the pipe cleaner at approximately every ¼-inch spot, back and forth. Glue one of each color pipe cleaner into each small circle.

31. Write on the poster to the left of this fourth circle (or you can type, print, cut out, and glue):

    Telophase

    1. The chromosomes uncoil.

    2. Two new nuclear membranes form.

Chapter 8: Lab *continued*

Example Poster 1:

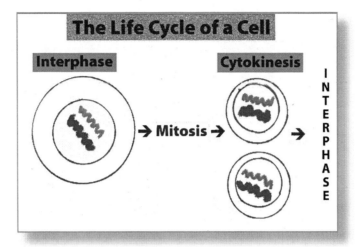

Example Poster 2:

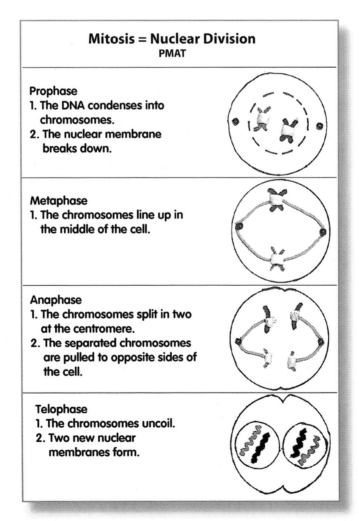

Mitosis with a Microscope

Chapter 8: Microscope Lab

If you can see the nucleus of a cell with a microscope, you should be able to see the nucleus undergoing mitosis with a microscope. If you want to be sure you will see mitosis, a good place to look would be something that grows quickly and easily. Have you ever had onions in your pantry start growing green shoots out of their tops? They grow quickly and easily.

Today you will look at a prepared slide of an onion tip undergoing mitosis. A prepared slide is one that has been prepared for you. The person preparing this slide made a thin slice of an onion tip, stained it with methylene blue, and fixed the slide cover on with a preservative so the slide would last a long time.

Materials

- Microscope

- Prepared slide of an allium (onion) root tip, l.s. (longitudinal section)

Procedure

1. Put the slide on the stage and focus it. Start at the lowest power and increase to 400x. Make sure you are at the tip end of the specimen since this is the most active site of mitosis.

2. Find and draw a view of each phase of mitosis on the view sheet.

Mitosis with a Microscope

Chapter 8: Microscope View Sheet

Name_____ **Date**_____

Specimen _____ Magnification _____

prophase

metaphase

anaphase

telophase

cytokinesis

Comments:

Absorb

Stem Cells

Chapter 8: Famous Science Series

You start as one cell. By the time you are born, that one cell has divided many times. Each cell division results in genetically identical cells. So how do you end up with over 200 different types of specialized cells? Stem cells, that's how.

Stem cells give rise to all the types of specialized cells in your body. They are able to differentiate into specialized cells. Stem cells can also renew themselves almost indefinitely, even after long periods of dormancy.

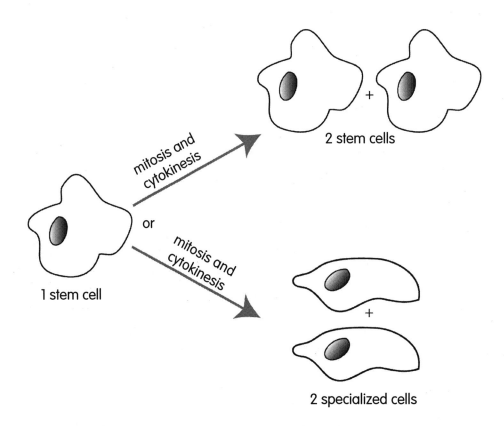

Pandia PRESS

Chapter 8: Famous Science Series *continued*

Research stem cells and answer the following:

1. The first cell that becomes a multicellular organism is an embryonic stem cell. It divides, becoming more stem cells. How do a small amount of embryonic stem cells become all the specialized cells that make a human?

2. What is the purpose of adult stem cells?

3. Stem cells are being used to treat cancer of the blood cells, called leukemia. How?

4. What other diseases do researchers hope stem cells can cure?

Mitosis–One Makes Two
Chapter 8: Show What You Know

Multiple Choice

1. Chromosomes are held together by

   ○ chromatids.
   ○ centromeres.
   ○ cytoplasm.
   ○ RNA.

2. Most of the cell cycle is spent in this phase:

   ○ mitosis
   ○ anaphase
   ○ prophase
   ○ interphase

3. The phases of mitosis occur in the following order:

   ○ telophase, anaphase, metaphase, prophase
   ○ anaphase, prophase, telophase, metaphase
   ○ metaphase, anaphase, prophase, telophase,
   ○ prophase, metaphase, anaphase, telophase

4. When DNA copies itself to make two genetically identical chromatids, it is called

   ○ duplication.
   ○ transcription.
   ○ replication.
   ○ translation.

5. The two genetically identical copies of a chromosome that are connected by a centromere are called

   ○ genes.
   ○ spindle fibers.
   ○ chromatids.
   ○ daughter cells.

6. Most bacteria reproduce asexually using the following process:

   ○ binary fission
   ○ mitosis
   ○ regeneration
   ○ budding

Chapter 8: Show What You Know *continued*

Questions

7. What are the three main stages of the cell cycle?

    1.

    2.

    3.

8. Interphase is divided into three parts. What happens in each part?

    G1 Phase:

    S Phase:

    G2 Phase:

9. Circle each correct answer. Humans are (haploid,   diploid) organisms. This means their ploidy is (1,   2). Humans have (23,   46) chromosomes in their somatic cells. The chromosomes are (single, in pairs). Humans get (all, half) their chromosomes from their mother.

10. A somatic cell with 8 chromosomes divides. How many daughter cells are produced? And how many chromosomes are in each daughter cell?

11. The daughter cells are _____ identical to each other with the same number of _____ in their cells.

Chapter 8: Show What You Know *continued*

12. What are the phases of mitosis this cell is in? Write the answer next to the cell. Number the phases according to the correct order they occur.

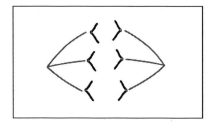

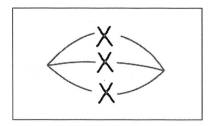

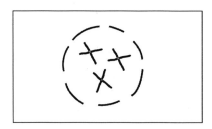

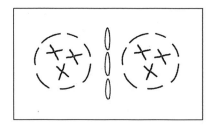

What type of eukaryotic cell is this, plant or animal? How do you know?

Pandia PRESS

Chapter 9: Meiosis Divides by Two and Makes You

Explore

Meiosis Flip Book

Chapter 9: Lab

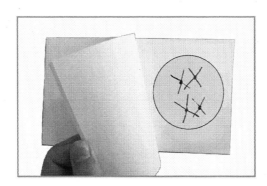

Materials

- Meiosis illustrations from your textbook
- Various colored pens or pencils
- Scissors
- Stapler

Procedure

1. Using the meiosis illustrations found in your textbook Chapter 9 Lesson as your guide, draw and color the following on the Meiosis Flip Book lab pages:

    a. The chromosomes in the correct position. Draw 2 homologous chromosomes.

    b. Nuclear membrane where there is one. Use a dashed or solid line, depending on the stage of meiosis.

2. When the drawings for each stage of meiosis are complete, cut the rectangles apart along the solid lines. Create a cover page for your flip book with one of the blank rectangles.

3. Stack the rectangles in order, then staple along the left-hand side to complete your flip book.

4. Teach the process of meiosis to an adult or fellow student using your flip book.

Meiosis Flip Book

Chapter 9: Lab Sheet pg. 1

Name_____ Date_____

Meiosis Flip Book pages

Prophase I	Telophase I
Metaphase I	Cytokinesis I
Anaphase I	

Meiosis Flip Book

Chapter 9: Lab Sheet pg. 2

Name_____ Date_____

Meiosis Flip Book pages

Prophase II	Telophase II
Metaphase II	Cytokinesis II
Anaphase II	

Explore: Meiosis and the Microscope

Chapter 9: Microscope Lab

Cells undergo mitosis all over a multicellular organism's body. Cells undergo meiosis, only in specific locations in an organism's body. Microscope Lab 9 uses another prepared slide, this time to look at the process of meiosis. The specimen is from the anther of a lily. An **anther** is where some types of plants make sperm, called **pollen**.

Materials

- Microscope
- Completed Meiosis Flip Book from Lab 9

- Prepared slide of a *Lilium* (lily), anther meiosis

Procedure

Center and focus your slide at 40x magnification. Then look at the slide with 100x and 400x magnification. Find cells undergoing each stage of meiosis. Match them with your completed Meiosis Flip Book. Put a small check or mark in your flip book for each phase of meiosis that you are able to identify on your slide.

Absorb

Down Syndrome
Chapter 9: Famous Science Series

At the start of meiosis, humans have 46 chromosomes. If you unraveled the DNA on your 46 chromosomes and stretched it out to see how long it was, it would reach from Earth to the moon! Can you imagine how easy it would be to tangle up a thin string that long? If you have ever gone fishing, you know how easy that would be.

Most of the time during cell division, the chromosomes separate without any issues. Sometimes though, issues happen. A permanent alteration in a gene's DNA sequence is called a **mutation**. Mutations are necessary for evolution and diversity amongst organisms. Mutations are also the cause of genetic disorders.

One type of mutation is called **trisomy**. The chromosomes condensing before mitosis and meiosis helps to separate the long, delicate strands that make chromosomes. Sometimes homologous chromosomes stick together instead of separating. When this happens, the gamete has two copies of the chromosome instead of one. Down syndrome, sometimes spelled Down's syndrome, is a genetic disorder caused by trisomy.

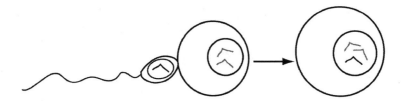

The male gamete, the sperm, has one copy of the chromosome but the female gamete, the egg, has two copies of the chromosome.

The resulting zygote has three copies of the chromosome. This is called trisomy.

Pandia PRESS

Chapter 9: Famous Science Series *continued*

What does the term *trisomy* mean?

People with Down syndrome have a trisomy at which chromosome number?

How many chromosomes total are in the somatic cells of a person with Down syndrome?

What are possible health or developmental impacts for people with Down syndrome?

Teddy McMahon

**People with Down syndrome can lead happy and productive lives.
Teddy McMahon enjoys playing the drums, and he likes having a
drawing of himself in a book.**

Meiosis Divides by Two and Makes You
Chapter 9: Show What You Know

1. Match the word with the best definition.

meiosis ◯   ◯ haploid cells

gametes ◯   ◯ XX

haploid ◯   ◯ cell division that creates gametes

diploid ◯   ◯ diploid cell made through the process of fertilization

fertilization ◯   ◯ XY

zygote ◯   ◯ chromosome number = n

male ◯   ◯ chromosome number = 2n

female ◯   ◯ 2 haploid cells fuse to make 1 diploid cell

Pandia PRESS

Chapter 9: Show What You Know *continued*

2. Fill in the blanks. Use each word in the word box only once.

half	gametes	variability	n	Y
number	genetically	mitosis	23rd	identical
X	2n	father	fertilization	
one	genes	X	zygote	
homologous	homologous	half	cytokinesis	

Mitosis results in _____ identical cells. The exact same _____ and _____ of chromosomes are in the daughter cells as was in the parent cell.

Meiosis results in genetic _____. The daughter cells are not genetically _____ to the parent cell. The chromosome number has been reduced by _____, which means only _____ the parent's chromosomes are in the daughter cells.

At the start of meiosis I, the parent cell has _____ chromosomes. During meiosis, the chromosome number is reduced to _____. The cells that are created during meiosis and cytokinesis are called _____.

During _____, the two gamete cells (one from a male and one from a female) fuse, making _____ cell, called a _____. In multicellular organisms, the unicellular zygote divides using the process of _____ and _____.

The _____ chromosome pair in humans are called the sex chromosomes. In females, this pair is a _____ pair. Females have two _____ chromosomes. In males, this pair is not a _____ pair of chromosomes. Males have one _____ and one _____ chromosome. The gender of a person is determined by their _____.

Chapter 10: Your Inheritance

What Makes You You?

Chapter 10: Lesson Activity

The following coloring activity is found in Chapter 10 Lesson in your Textbook.

Homologous Chromosomes

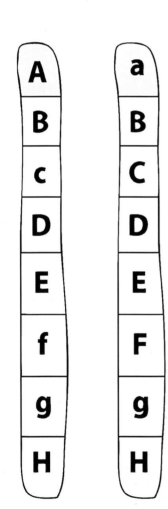

1. Each segment is a gene. The genes come in pairs, one on each homologous chromosome.
2. The set of all the genes in an organism is their genotype. An organism's genotype has all their alleles.
3. Phenotype is the appearance and chemical makeup of an organism. Brown eyes and sickle cell anemia are both examples of phenotype. An organism's phenotype is determined by its genotype.
4. Letters are used for alleles. The same letters are used for alleles of the same gene. Different letters are used for different genes. A and a are alleles of each other but B is not. Color the alleles the same color, different from the others. For example, A and a are the same color as each other but not the same color as B and B, which are the same color as each other.
5. When two alleles are written exactly the same, the organism is homozygous for that allele. The chromosome pair is homozygous for the BB alleles. Write "ho" between the homozygous alleles.
6. When the two alleles are written differently, the organism is heterozygous for that allele. This chromosome pair is heterozygous for the Aa alleles. Write "hz" between the heterozygous alleles.
7. Alleles can be dominant or recessive. Uppercase letters are used to show an allele is dominant. Lowercase letters are used to show an allele is recessive.
8. A dominant allele is one that is expressed in the phenotype even if there is only one copy of it present in the genotype. Write a "D" next to hz or ho, if a dominant allele is in the pair.
9. A recessive allele is one that is expressed in the phenotype only if there are two copies of it present in the genotype. Write an "r" next to "ho" if there is a pair of recessive alleles in the pair.

Explore Family Traits
Chapter 10: Lab

Blood-related members of a family inherit a unique set of genes. Members of the same family also share some genes. For this lab, you will be filling out a questionnaire for yourself and your family members. If you are adopted or otherwise don't have access to your own blood-related family members, then complete this lab with another group of blood-related individuals choosing one member to represent "Me" on the lab sheets. You are going to interview members of the family, asking them about their phenotype for different traits. Using the phenotype, you will be able to determine the genotypes of the family for these traits. In the case of a dominant trait that is present in the family, you might only be able to determine that the family members are heterozygous dominant or homozygous dominant for the trait. In some cases an exact genotype can be determined using family members. For example, a non-tongue-rolling mother and a tongue-rolling father can only have a non-tongue-rolling child *if* the father was heterozygous for this trait because tongue-rolling is dominant over non-tongue-rolling. The father must have passed his allele for non-tongue-rolling to his offspring.

Materials

- Blood-related family members to interview (The more the better. There's space on the chart for up to 6 family members including "Me.")

Procedure

1. Fill in the questionnaire for yourself and several of your blood-related relatives, or for another group of blood-related members if you are unable to interview yours. The more relatives you interview, the more accurate the results will be.

2. When you have finished the questionnaire, complete the Who Shares the Good Looks? worksheet. For each relative, write their relationship to you (or to the person you've chosen to be the "Me" in this lab), e.g., mother, grandfather, sister, etc. For each phenotype, write each relatives' genotype. For example, if you have light blue eyes and your mother has brown eyes, write eeeeeeee for your eye color genotype, and EEEEEEEe for your mother's.

3. Look at the traits you have listed as either homozygous dominant or heterozygous recessive on the worksheet. Do you have enough information to determine whether you or your relative is one or the other? For example, if you can't roll your tongue but both your parents can, then you will know that they each carry the recessive allele. So their genotype must be Rr in order for yours to be rr.

Pandia PRESS

Chapter 10: Lab *continued*

4. Check the box for each genotype that matches yours for dominant or recessive. These are the relatives who share traits with you, and in some cases where you got your good looks. For example:

	Me	**Relative 1**	**Relative 2**	**Relative 3**	**Relative 4**	**Relative 5**
	Jake	*Mom*	*Mike*	*Uncle Terry*	*Grace*	*Dad*
Relationship	*Myself*	*Mother*	*Cousin (Dad's side)*	*Mom's brother*	*Sister*	*Father*

Phenotype	**Genotype**					
Tongue Rolling	*rr* ☐	*rr* ☑	*Rr or RR* ☐	*rr* ☑	*Rr* ☐	*Rr* ☐

Phenotype Descriptions and Instructions

Blood Type. You inherit the type of blood you have from your mother and father. Here is a Punnett square showing how it works:

blood type allele	A	B	O
A	AA	AB	AO
B	BA	BB	BO
O	OA	OB	OO

AA = blood type A

AB and BA = blood type AB

AO and OA = blood type A

BB = blood type B

BO and OB = blood type B

OO = blood type O

Eye Color. The genes that code for eye color are on four different chromosomes. Therefore there are eight alleles for eye color (4 x 2). These are represented by E and e on your questionnaire. The number of genes that code for eye color leads to the amount of variability of eye color found in humans. The eye colors on your questionnaire are listed in order of their dominance, starting with black as the most dominant and ending with blue as the most recessive.

Tongue Rolling. Can you roll your tongue into a tube shape? The ability to tongue-roll is controlled by a dominant allele, R. People who are tongue-rollers have at least one copy of the dominant, R, allele. They are either homozygous dominant, RR, or heterozygous dominant, Rr, for this trait. People who are not tongue rollers are homozygous recessive, rr, for this trait.

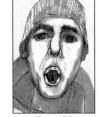

Rr or RR

Widow's Peak. Push your hair back off your forehead and look in a mirror. If your hairline has a V at the top of your forehead, you have a widow's peak. People with a widow's peak have the dominant phenotype for this trait. People with a straight hairline have the recessive phenotype for this trait.

Ww or WW

Chapter 10: Lab *continued*

Freckles. Freckles are controlled by the dominant gene F. If you do not have freckles, you are homozygous recessive for them, ff. (Sun freckles do not count.)

Dimples. Do you have dimples? They may be cute but a dimple is actually an anomaly of face muscle that causes a dent in the cheek, especially when smiling. Dimples are a dominant trait (Dd or DD).

Detached Earlobes. Are you lobeless or do you have earlobes? If your earlobes are detached from the side of your face, you have free ear lobes, which is the dominant trait. If your earlobes are attached to the side of your face, you are lobeless, and are homozygous recessive for this trait, ee.

ee   Ee or EE

Hair Color. Hair color, like eye color, comes over a broad range. Hair color is controlled by three different chromosomes (six alleles). The alleles for dark hair color are dominant over the alleles for light hair color. The lighter someone's hair is, the more recessive hair color alleles that person has.

Hair Texture. The dominant phenotype for hair texture is straight. People with curly hair have the recessive phenotype.

Wears Glasses. Good eyesight is dominant over impaired eyesight. If you wear glasses or contacts, you have the recessive phenotype.

Straight Teeth. Straight teeth are dominant over crooked teeth. Make sure and ask your older relatives if they have always had straight teeth; braces have been around a long time.

Hairy Fingers. Your fingers have three segments. Look at the middle segment to see if hair is growing on it. Hair growth on this part of your finger is a dominant phenotype.

Short Big Toe. Is your big toe shorter than the second toe that is right next to it? A longer second toe is a dominant phenotype. If the big toe is longer or the same length as the second toe, you have the recessive phenotype.

Vulcan Sign. Can you make the Vulcan sign (pictured at the right)? If you can, you and Mr. Spock have something in common. You both have the dominant phenotype for this trait.

Vv or VV

Top Thumb. Clasp your hands together. Which thumb is on top, on the outside closest to you? The dominant phenotype is to have the left thumb on top. I bet you didn't know that was an inherited trait!

Pandia PRESS

Family Traits Questionnaire

Chapter 10: Lab Sheet

Name_____ Date_____

Phenotype	Genotype	Me	Relative 1	Relative 2	Relative 3	Relative 4	Relative 5
Blood Type		A	A	A	A	A	A
		B	B	B	B	B	B
		O	O	O	O	O	O
		AB	AB	AB	AB	AB	AB
Eye Color	EEEEEEEE	black	black	black	black	black	black
	EEEEEEEe	brown	brown	brown	brown	brown	brown
	EEEEEEee	hazel	hazel	hazel	hazel	hazel	hazel
	EEEEEeee	gray	gray	gray	gray	gray	gray
	EEEEeeee	amber	amber	amber	amber	amber	amber
	EEEeeeee	blue green	blue green	blue green	blue green	blue green	blue green
	EEeeeeee	green	green	green	green	green	green
	Eeeeeeee	dark blue	dark blue	dark blue	dark blue	dark blue	dark blue
	eeeeeeee	light blue	light blue	light blue	light blue	light blue	light blue
Tongue Rolling	Rr, RR	yes	yes	yes	yes	yes	yes
	rr	no	no	no	no	no	no
Widow's Peak	WW, Ww	yes	yes	yes	yes	yes	yes
	ww	no	no	no	no	no	no
Freckles	FF, Ff	yes	yes	yes	yes	yes	yes
	ff	no	no	no	no	no	no
Dimples	DD, Dd	yes	yes	yes	yes	yes	yes
	dd	no	no	no	no	no	no

Chapter 10: Lab Sheet *continued*

Phenotype	Genotype	Me	Relative 1	Relative 2	Relative 3	Relative 4	Relative 5
Detached Earlobes	EE, Ee	yes	yes	yes	yes	yes	yes
	ee	no	no	no	no	no	no
Hair Color	HHHHHH	black	black	black	black	black	black
	HHHHHh	brown	brown	brown	brown	brown	brown
	HHHHhh	light brown	light brown	light brown	light brown	light brown	light brown
	HHHhhh	auburn	auburn	auburn	auburn	auburn	auburn
	HHhhhh	red	red	red	red	red	red
	Hhhhhh	dark blond	dark blond	dark blond	dark blond	dark blond	dark blond
	hhhhhh	blond	blond	blond	blonde	blonde	blonde
Hair Texture	AA, Aa	straight	straight	straight	straight	straight	straight
	aa	curly	curly	curly	curly	curly	curly
Wears Glasses	GG, Gg	no	no	no	no	no	no
	gg	yes	yes	yes	yes	yes	yes
Straight Teeth	TT, Tt	yes	yes	yes	yes	yes	yes
	tt	no	no	no	no	no	no
Hairy Fingers	HH, Hh	yes	yes	yes	yes	yes	yes
	hh	no	no	no	no	no	no
Short Big Toe	BB, Bb	yes	yes	yes	yes	yes	yes
	bb	no	no	no	no	no	no
Vulcan Sign	VV, Vv	yes	yes	yes	yes	yes	yes
	vv	no	no	no	no	no	no
Top Thumb	TT, Tt	left	left	left	left	left	left
	tt	right	right	right	right	right	right

Who Shares the Good Looks?

Chapter 10: Lab Sheet

Name: _____ Date: _____

	Me	Relative 1	Relative 2	Relative 3	Relative 4	Relative 5
Relationship						

Phenotype	Genotype					
Blood Type						
Eye Color						
Tongue Rolling						
Widow's Peak						
Freckles						
Dimples						
Detached Earlobes						
Hair Color						
Hair Texture						
Wears Glasses						
Straight Teeth						
Hairy Fingers						
Short Big Toe						
Vulcan Sign						
Top Thumb						

Pandia PRESS

Phenotype Under the Scope
Chapter 10: Microscope Lab

What do you think different-colored strands of hair will look like under the microscope? Will you be able to see why they are differently colored from one another? Today you will find out. You need to collect strands of human hair in as many colors as possible. The hairs do not need to be long, just different colors. Try to get strands of hair that have not been artificially colored. You might include one strand that has been dyed and one that has been bleached, but you do not need more than that. This lab looks at phenotype. When you are looking at chemically processed hair, you are not seeing phenotype; you are looking at a chemical process. Try to get two hairs from someone who is going gray. Get one hair from your graying subject that has not turned silver yet and one that has, examine the difference.

Materials

- Hair strands in different colors
- Slides (same number as strands of hair)
- Slide covers
- Scissors
- Tape

- Syringe
- Water
- Metric ruler
- Microscope

Procedure

1. Cut one strand of hair 3 cm long. Lay the cut piece of hair across a slide lengthwise. Tape the very ends of the hair with two small pieces of tape. Drip water along the hair. Cover with a slide cover. Look at it under the microscope. Leave this slide intact as a reference.

2. Repeat the procedure for each piece of hair, mounting each on its own slide.

3. Draw a view of a strand of your own hair at your favorite magnification on the Microscope Lab Sheet.

Pandia PRESS

Phenotype Under the Scope

Chapter 10: Microscope Lab Sheet

Name_____ Date_____

Specimen_____ Type of Microscope _____

Type of Stain_____ Magnification _____

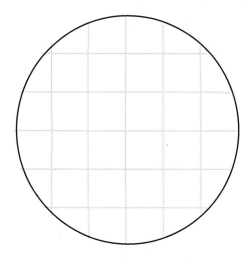

1. Describe what makes hair different in color.

2. What did you notice about the gray versus the colored strand of hair from the same person?

3. If you include bleached and/or dyed hair, what did you notice different about it?

Additional Comments:

Gregor Mendel

Chapter 10: Famous Science Series

Gregor Mendel

Father of Modern Genetics

When and where was Gregor Mendel born?

What did Mendel do so that he could continue his education?

What is the blending theory of inheritance?

From 1856 to 1863, Mendel conducted an experiment with over 28,000 plants. What type of plant did he use?

What seven traits did he study in these plants?

    1.

    2.

    3.

Pandia PRESS

Chapter 10: Famous Science Series *continued*

    4.

    5.

    6.

    7.

Did Mendel prove or disprove the blending theory of inheritance?

What two laws did Mendel discover?

    1.

    2.

When Mendel crossed true-breeding green peas and white peas, he got all green peas in the F1 generation. When he crossed two of the green peas from the F1 generation he got ¾ green peas and ¼ yellow peas in the F2 generation. Which is the dominant trait and which is the recessive trait?

Was Mendel famous in his lifetime?

Make Your Own Qwitekutesnute
Chapter 10: Activity

Gregor Mendel formulated two laws of science based on his research data: the Law of Segregation and the Law of Independent Assortment. The **Law of Segregation** states that allele pairs separate (segregate) during meiosis when gametes form. When a cell goes from being diploid, with two sets of paired homologous chromosomes, to haploid, with only one set of chromosomes, the allele pairs on each chromosome separates. The **Law of Independent Assortment** states that allele pairs assort independently of one another during gamete formation.

Even organisms from the Island of Mythical Creatures reproduce using Mendelian genetics. When qwitekutesnutes have offspring, the allele pairs of the parents separate, following the Law of Segregation. The allele pairs separate independently of one another, following the Law of Independent Assortment. When the alleles from the mother and father pair during fertilization, the dominant phenotype will be expressed even if only one copy of the dominant allele is present.

Qwitekutesnutes have 12 different traits that have been identified (unless you add one more that you have identified during the course of your research). You are going to cut (segregate) the allele pairs of each parent found on the Qwitekutesnute Traits sheet into separate cards. You will flip a coin to determine independently which allele from each parent the offspring receives. You will do the same to determine the gender of your qwitekutesnute.

Materials

- Scissors
- Coin

- Colored pencils or markers

Chapter 10: Activity *continued*

Procedure

1. Both Qwitekutesnute Traits sheets have 12 traits listed, plus the sex chromosomes. There is one blank row of squares at the bottom of each sheet, so that you can add traits of your own if you want. If you add traits, choose one choice for that trait and write them into the squares next to each other. If you add traits, decide which allele is dominant and which is recessive.

2. For both sheets, cut out the first row of rectangles. Turn them over and put them next to each other. Make sure you keep the male and female cards separate. Cut out the second row of squares and turn them over. Put them next to each other below the first set of squares. Do this for all the traits and the gender-determination squares.

3. Go down the rows choosing the traits. You do this by flipping the coin. Heads means the trait on the top is the one your qwitekutesnute has. Tails means the trait on the bottom is the one your qwitekutesnute has. Do this for both the male and female cards. As you choose each of one of the two choices, put it on the qwitekutesnute template. You will have one choice from the male and one from the female for a total of two squares for each trait. You will draw the phenotype for that trait on the Qwitekutesnute Template.

4. Each trait has a dominant and a recessive allele. Look for a D or an r written in the bottom right hand corner of each card. If you choose one to two cards with the dominant allele, the qwitekutesnute has the dominant trait. If you choose two recessive alleles, the qwitekutesnute is homozygous recessive for the trait and has the recessive trait.

5. Draw each trait in the following order. The choices for each trait are:

   ○ Small-body size (3 kilograms) or large-body size (5 kilograms). Choose the correct body-size template, large-bodied or small-bodied.

   ○ Big fluffy tail or a thin short tail, add the tail to your qwitekutesnute.

   ○ Stars or stripes, qwitekutesnutes come in two fur patterns. Draw the fur pattern on the template.

   ○ Spiked hair or ear tufts, qwitekutesnutes have two possible hairstyles that are genetically controlled. Draw the hairstyle on the template

   ○ A male qwitekutesnute has an X and a Y chromosome. A female qwitekutesnute has two X chromosomes. You cannot tell a male and female qwitekutesnute apart by looking at them. Male qwitekutesnutes have boy names and female qwitekutesnutes have girl names. Choose a name for your qwitekutesnute based on its gender.

   ○ Almond-shaped eyes or round-shaped eyes, qwitekutesnutes have almond-shaped eyes or round-shaped eyes. Draw eyes in the appropriate shape on the face of your qwitekutesnute.

Chapter 10: Activity *continued*

○ Gray eyes or green eyes, color the eyes of the qwitekutesnute either green or gray.

○ 3 toes or 5 toes on their feet, draw either three toes or five toes on each of your qwitekutesnutes' feet.

○ 3 fingers or 4 fingers on their hands, draw either three fingers or four fingers on each of your qwitekutesnutes' hands.

○ Have a cry that goes Kee-Kee, or one that goes Koo-Koo. Your qwitekutesnute is probably calling to you, telling you it is hungry. They are ALWAYS hungry. Draw a carton bubble above your qwitekutesnute's head and write in the call your qwitekutesnute makes.

○ Bushy eyebrows or no eyebrows, if your qwitekutesnute has bushy eyebrows, draw them over its eyes. If it has no eyebrows, do not draw any.

○ Qwitekutesnutes can be dark gray, light gray, or white. This is an example of incomplete dominance; neither dark gray nor white alleles are dominant over each other. When a qwitekutesnute is heterozygous for fur color, the two colors mix and they are light gray.

  • 2 dark gray alleles make a dark gray qwitekutesnute

  • 1 dark gray + 1 white allele make a light gray qwitekutesnute

  • 2 white alleles make a white qwitekutesnute

○ 4 whiskers on each cheek or 7 whiskers on each cheek, draw either four whiskers or seven whiskers on each side of your qwitekutesnute's face.

○ Finish with any traits you added.

○ I would love to see your qwitekutesnutes! If you can, please email an image of them to blair@pandiapress.com. (Disclaimer: Your pictures could be used online or in a publication.)

Pandia PRESS

Make Your Own Qwitekutesnute

Chapter 10: Activity Sheet

Female Qwitekutesnute Traits

small-body size, 3 kg _r	large-body size, 5 kg _D
big fluffy tail _r	thin short tail _D
stars _D	stripes _r
spiked hair _D	ear tufts _r
X	X
almond-shaped eyes _r	round-shaped eyes _D
gray eyes _D	green eyes _r
3 toes _r	5 toes _D
3 fingers _D	4 fingers _r
says Kee-Kee _D	says Koo-Koo _r
bushy eyebrows _r	no eyebrows _D
dark gray	white
4 whiskers _r	7 whiskers _D

Chapter 10: Activity Sheet *continued*

Male Qwitekutesnute Traits

small-body size, 3 kg _r	large-body size, 5 kg _D
big fluffy tail _r	thin short tail _D
stars _D	stripes _r
spiked hair _D	ear tufts _r
Y	X
almond-shaped eyes _r	round-shaped eyes _D
gray eyes _D	green eyes _r
3 toes _r	5 toes _D
3 fingers _D	4 fingers _r
says Kee-Kee _D	says Koo-Koo _r
bushy eyebrows _r	no eyebrows _D
dark gray	white
4 whiskers _r	7 whiskers _D

Chapter 10: Activity Sheet *continued*

Large-Bodied Qwitekutesnute

Chapter 10: Activity Sheet *continued*

Small-Bodied Qwitekutesnute

PandiaPRESS

Inheritance

Chapter 10: Show What You Know

1. Match the word with the best definition.

allele ⬭   ⬭ the set of genes in an organism

homologous chromosomes ⬭   ⬭ an allele that is expressed only if two copies are
                                      present in the genotype

genotype ⬭   ⬭ a chart used to predict genotype based on the
                      parents' alleles

phenotype ⬭   ⬭ forms of a gene

dominant allele ⬭   ⬭ two of the same copies of an allele

recessive allele ⬭   ⬭ an organism's appearance

homozygous ⬭   ⬭ chromosome pairs

heterozygous ⬭   ⬭ an allele that is expressed if one or more copies
                          are present in the genotype

Punnett square ⬭   ⬭ two different copies of an allele

Chapter 10: Show What You Know *continued*

2. Complete the Punnett Square.

Qwitekutesnutes can have either a spiked hairdo or flat hairdo with elaborate ear tufts. One qwitekutesnute parent has a spiked hairdo (Hh), and one has ear tufts (hh). With these two parents, what is the probability of a qwitekutesnute with ear tufts? Use the Punnett square to determine this, and then transfer the information to Table 1 and answer the questions.

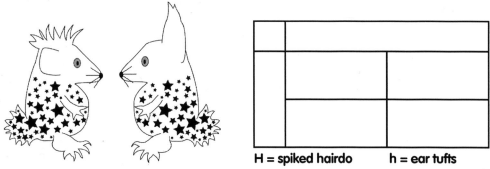

H = spiked hairdo   h = ear tufts

Table 1

Genotype	Genotype Probability	Genotype Fraction	Genotype Percentage	Phenotype	Phenotype Probability	Phenotype Fraction	Phenotype Percentage

What is the probability of a qwitekutesnute baby from this pair having ear tufts?

If the qwitekutesnute parents have 12 babies, how many should have ear tufts? Will that many definitely have ear tufts?

If qwitekutesnute parents both have gray eyes (a dominant trait among qwitekutesnutes), could they have green-eyed offspring (a recessive trait)? Explain your answer.

If qwitekutesnute parents both have 4 whiskers, a recessive trait, could they have offspring with 7 whiskers? Explain your answer.

Pandia PRESS

Chapter 10: Show What You Know *continued*

Multiple Choice

3. Law of Segregation states

    ○ alleles for different traits don't mix

    ○ allele pairs separate during meiosis

    ○ different species can't reproduce
       with each other

    ○ cell segregate after cytokinesis

4. Law of Independent Assortment states

    ○ allele pairs assort independently of
       one another

    ○ allele pairs separate during meiosis

    ○ different species can't reproduce
       with each other

    ○ cells segregate after cytokinesis

5. A scar on your chin is an example of

    ○ dominant alleles

    ○ recessive alleles

    ○ genotype

    ○ phenotype

6. The allele pair Ww is

    ○ homozygous dominant

    ○ homozygous recessive

    ○ heterozygous

7. The allele pair BB is

    ○ homozygous dominant

    ○ homozygous recessive

    ○ heterozygous

8. The allele pair ee is

    ○ homozygous dominant

    ○ homozygous recessive

    ○ heterozygous

9. If two parents with brown hair have a baby with
    blond hair, the allele for blond hair must be

    ○ recessive

    ○ dominant

    ○ there is not enough information

    ○ heterozygous

10. Traits are

    ○ determined by only genotype

    ○ determined by phenotype

    ○ inherited and acquired

    ○ your alleles

11. Your genotype is

    ○ different for different specialized cells

    ○ the set of genes in the somatic cells in your
       body

    ○ the same in gametes and somatic cells

12. Your traits are your

    ○ phenotype

    ○ alleles

    ○ genotype

    ○ ploidy

Chapter 10: Show What You Know *continued*

13. **Extra Practice.** As you know, qwitekutesnutes can have a spiked hair or ear tufts. The allele for spiked hair, H, is dominant over ear tufts, h. Below are six Punnett squares, with possible crosses for qwitekutesnutes hair. When you complete the Punnett squares, write the possible percentage of each phenotype beside the Punnett square.

HH x HH

phenotype percentage:

HH x Hh

phenotype percentage:

HH x hh

phenotype percentage:

Hh x Hh

phenotype percentage:

Hh x hh

phenotype percentage:

hh x hh

phenotype percentage:

Inside View of a Frog
Chapter 11: Dissection Lab

A dissection is good way to look at how an organism is put together. Today you will look at the component parts of a frog. Frogs have cells that make tissues. They have different types of tissues that make organs. The organs in frogs group together into organ systems. The different organ systems make an organism, in this case a frog. First, you will examine the external anatomy of the frog. Next, you will examine inside the mouth and its passages. Then, you will cut the frog open and examine its internal anatomy.

Materials

- Preserved frog
- Safety goggles
- 6 to 10 dissecting pins
- Dissecting tray
- Paper towels
- Gloves
- Forceps
- Scissors

- Tape measure with cm
- Slides
- Water
- Dissecting probe or tweezers
- Medicine dropper
- Baggies for disposal and storage
- X-Acto knife

Optional/helpful items:
- Scalpel
- Apron
- Magnifying glass

Procedure

1. Put your lab sheets and pencil next to where you are going to perform the dissection. You will fill these out as you perform the dissection.

2. Put on the safety goggles, gloves, and apron.

3. Take the frog out of its container, rinse it off, and place it on the dissecting tray.

4. Use the Dissection Lab Sheets as references as you dissect the frog.

Chapter 11: Dissection Lab *continued*

External Anatomy

1. Examine the frog. Look at it on its back and on its belly.

2. Answer the questions and fill in the blanks on the External Anatomy Lab Sheet.

The Mouth

1. Put the frog on its top side, belly up. Using the scissors, carefully cut the sides of the mouth.

2. Turn the frog over.

3. Follow the instructions for examining inside the mouth.

4. Answer the questions and fill in the blanks on the Mouth Anatomy Lab Sheet.

Internal Anatomy: The Dissection Process

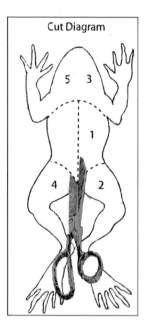

Cut Diagram

1. Put the frog on its top side, belly up, on the dissecting tray.

2. Pin the frog for dissection. Put one pin through each leg.

3. With your scissors or a scalpel, make a small cut through the belly of the frog. Do this in the crook where the two back legs come together. BE VERY CAREFUL YOU ONLY CUT THROUGH THE SKIN!

4. With the scissors, CAREFULLY cut the skin from this point up the center of the frog to its neck.

5. Make cuts near the arms and legs, so you can peel back the skin.

6. The skin flaps might need to be separated from the muscle below. If they do, carefully pull the skin with the forceps as you cut the connecting tissue with the X-Acto knife. Pin these flaps of skin to the tray.

7. Take your time looking at the muscle and the blood vessels. When you are done, you need to cut through the muscle. BE VERY CAREFUL YOU ONLY CUT THROUGH THE MUSCLE! Cut up the midline as you did for the skin.

8. Before cutting across the arm lines, you need to cut through the breastbone. Turn the scissor blades sideways, to prevent cutting the heart or internal organs.

9. Cut along the front and back legs and pin the muscle flaps open.

10. If the frog is a female, the body may be filled with eggs and an enlarged ovary. If they are, you will need to remove these carefully. Save them when you remove them. You can look at the frog eggs and ovary later with a magnifying glass.

Internal Anatomy: The Examination

Answer the questions and fill in the blanks on the Internal Anatomy Lab Sheet. The instructions for examining the organs are on the lab sheet.

Inside View of a Frog

Chapter 11: Dissection Lab Sheet

Name_____ **Date**_____

External Anatomy

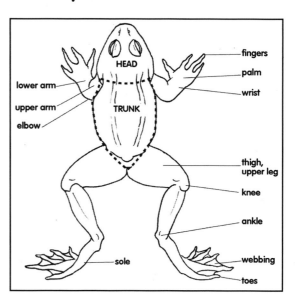

Check off the parts as you identify them on your frog:

○ **Head**

○ **Trunk**

○ **Belly**

○ **Eyes**

○ On the eyes, find the **Nictitating Membrane**. The upper and lower lids of the eyes do not close. The eyes are covered with a nictitating membrane. The nictitating membrane allows the frog to see underwater and protects the frog's eyes from drying out on land.

○ **Tympanum**. The tympanum is behind the eye. This is the frog's hearing organ.

○ **Nares**. The nares are the nostrils of the frog.

○ **Mouth**

On the front legs, find:

○ **Fingers**

○ **Palm**

○ **Wrist**

○ **Elbow**

On the back legs, find:

○ **Thigh**

○ **Knee**

○ **Ankle**

○ **Webbing**

○ **Toes**

○ **Sole**

Pandia PRESS

Chapter 11: Dissection Lab Sheet *continued*

Questions to Answer

1. How many fingers does the frog have?

2. One method for determining gender in a frog is to examine the fingers on its hands. Male frogs have thick pads on their "thumbs." Female frogs do not. Does your frog have these pads? Do you think your frog is a male or a female?

3. How many toes does the frog have?

4. What is the length of the frog? (Measure in centimeters.)

5. Describe the color difference between the top of the frog and the belly of the frog. The color difference protects a frog from predators when it is in the water. How do you think this works?

6. A frog's nares are on the top of its nose. How would this help the frog when swimming?

7. How do you think a frog's webbed feet help it in the water?

8. The outer covering of the frog, its skin, is its largest organ. What do you think your largest organ is?

Pandia PRESS

Chapter 11: Dissection Lab Sheet *continued*

Mouth Anatomy

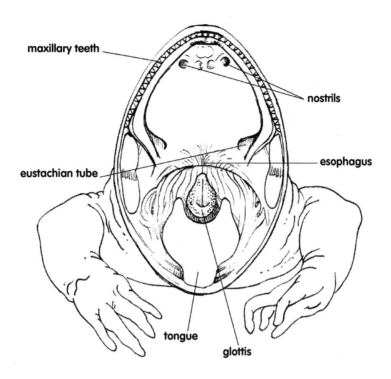

Check off the parts as you identify them on your frog:

○ **Tongue**

○ **Teeth**. Feel for the teeth inside the frog's mouth on the top and bottom jaw and between the internal nostrils.

○ **Internal Nostrils**. Put a probe into one of the external nostrils of the frog. Watch it come out of an internal nostril.

○ **Glottis**. Open the mouth wide and look in the back to find it. The glottis is an opening in the back of the throat; it is used for breathing and leads to the lungs. It is before the esophagus.

○ **Esophagus**. Also in the back of the throat, behind the glottis, is a larger opening called the esophagus. The esophagus is the tube that leads from the mouth to the stomach.

○ **Eustachian tubes**. There are two. They are above the esophagus.

Chapter 11: Dissection Lab Sheet *continued*

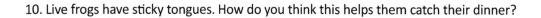

More Questions to Answer

9. What happened when you put a probe through the frog's nostril?

10. Live frogs have sticky tongues. How do you think this helps them catch their dinner?

11. Eustachian tubes equalize pressure in the ear. What external organ do you think they work with?

12. Think about experiences you have had with drainage between your mouth and nose. Do you think that your nasal passage goes directly into your mouth too?

Chapter 11: Dissection Lab Sheet *continued*

Internal Anatomy

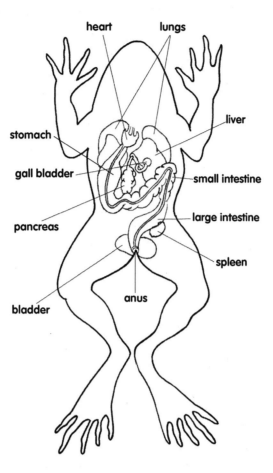

Check off the parts as you identify them on your frog:

○ **Liver**. The largest organ in the abdominal cavity. It is dark reddish-brown.

○ **Gallbladder**. CAREFULLY lift the liver out using the forceps. The greenish sac attached to the liver is the gallbladder. Bile is stored in the gallbladder. It is an organ that helps digest fat.

○ **Lungs**. They are two small sacs on either side of the midline, partly hidden by the liver. Insert a medicine dropper down the frog's glottis and inflate the lungs.

○ **Heart**. Between the lungs is the heart. There might be a filmy covering over the heart; if there is, lift it off with the forceps.

○ **Veins and arteries**. You will need to CAREFULLY lift the heart out. Veins carry blood to the heart. Arteries carry blood away from the heart.

○ **Stomach**. The stomach leads into the small intestine.

○ **Small intestine**. Attached to the stomach is the small intestine.

○ **Large Intestine**. The large intestine goes to the anus. With your finger, and without touching the frog, trace the path of food as it goes through the organs in the digestive system. Start with the mouth, to the stomach, then to the small and large intestines, and end with the anus.

(continued on next page)

Chapter 11: Dissection Lab Sheet *continued*

PandiaPRESS

○ **Anus**

○ **Pancreas**. Lift the small intestine to see the pancreas. It is a thin, ribbon-like structure between the small intestine and the stomach. The pancreas helps with digestion.

○ **Spleen**. Near the pancreas is a dark red, spherical organ, called the spleen. The spleen is a holding area for blood.

○ **Bladder**

Final Questions to Answer

13. Look at the blood vessels under the skin. What is their purpose?

14. The organs in the digestive system are the mouth, tongue, teeth, esophagus, stomach, small intestine, large intestine, and anus. If this frog ate a bug, write the path it would take.

PandiaPRESS

Alexander Fleming
Chapter 11: Famous Science Series

Infection, Antibiotics, and War

Alexander Fleming

You are out shopping. Without thinking, you put your hand on the checkout counter and then put lip balm on without washing your hands. There were live bacteria on the counter and now they are on your lips. You lick your lips, and now they are inside you. You are going to be sick. It will take a couple of days before you know it, but you are. First, your throat will hurt. Then, you will start out sneezing. Next, your whole body will start hurting and you will have a fever. Your parent will take you to the doctor, and the doctor will prescribe antibiotics. Think about it: no more bacteria than would fit on the tip of your finger got inside you and made your whole system sick.

1. When were antibiotics discovered? Who discovered them? What was the name of the first antibiotic?

2. This scientist fought in what war? What did he see in that war that caused him to dedicate his life to finding an antibacterial agent?

Chapter 11: Famous Science Series *continued*

3. For much of human history, there has been a war going on somewhere. What happened to wounded soldiers in the time before there were antibiotics? We can look at data from the Civil War to help answer this question. The Surgeon General of the United States, William A. Hammond of the Union Medical Corps, kept detailed records of the causes of deaths of Union soldiers. Based on these records:

   A. What percentage of Union soldiers who died, died from infectious disease?

   B. What is the estimated percentage of Confederate soldiers who died from infectious disease?

   C. Of the 620,000 soldiers who died in the Civil War, approximately how many died from infectious disease?

   Think of all the wars in history that occurred before World War II and the discovery of penicillin. Think of all the people in all the wars that died from infections that could have been cured with antibiotics. That is a lot of people! Isn't it?

Pandia PRESS

Multicellular Organisms
Chapter 11: Show What You Know

Multiple Choice

1. Your blood connects your entire body. Even though it is a liquid it is considered a type of this tissue:

   ○ Epithelial tissue
   ○ Connective tissue
   ○ Muscle tissue
   ○ Nerve tissue

2. Your brain has sensors going back and forth to it as it constantly monitors your environment. It relies on this type of tissue to transmit information:

   ○ Epithelial tissue
   ○ Connective tissue
   ○ Muscle tissue
   ○ Nerve tissue

3. Your body's largest organ is your skin. Its primary job is to protect the inside from the outside and to hold you together. Your skin is lined with this type of tissue:

   ○ Epithelial tissue
   ○ Connective tissue
   ○ Muscle tissue
   ○ Nerve tissue

4. Your heart beats your entire life. It never stops working. It relies on this tissue to keep it beating:

   ○ Epithelial tissue
   ○ Connective tissue
   ○ Muscle tissue
   ○ Nerve tissue

Chapter 11: Show What You Know *continued*

5. Matching. Match the vocabulary word with the definition.

cell ◯   ◯ connects and gives support

tissue ◯   ◯ the basic building block of a living organism

epithelial tissue ◯   ◯ a living being, what you get when you put all the organ systems together

connective tissue ◯   ◯ two or more types of specialized tissue that are working together

muscle tissue ◯   ◯ a group of two or more organs that work together to perform a function

nerve tissue ◯   ◯ lines the inside and outside of your body

organ ◯   ◯ a group of the same type of specialized cells

organ system ◯   ◯ generates and conducts electrical signals through your body

organism ◯   ◯ contracts and relaxes, it's found in your muscles

Pandia PRESS

Chapter 12: Plant Anatomy

Plants, Roots, and Shoots

Chapter 12: Lesson Activity

The following coloring activity is found in the Chapter 12 Lesson in your Textbook.

The Root System

The organs of the root system are the primary root, secondary root, root cap, and the root hairs. Below, read about each organ in the root system and color each organ as instructed.

- **Primary root**—The **primary root** is the thickest of the roots. It grows down. <u>Color it brown.</u>

- **Lateral roots**—The **lateral roots**, or secondary roots, grow from the primary root. They are thinner than the primary root. They grow laterally. <u>Color them brown.</u>

- **Root caps**—The **root caps** protect the root as it grows and pushes through the soil. <u>Color them dark drown.</u>

- **Root hairs**—**Root hairs** are thin hair-like roots that are one cell thick in diameter. Water and nutrients absorb through the root hairs into the roots. <u>Trace them over in black.</u>

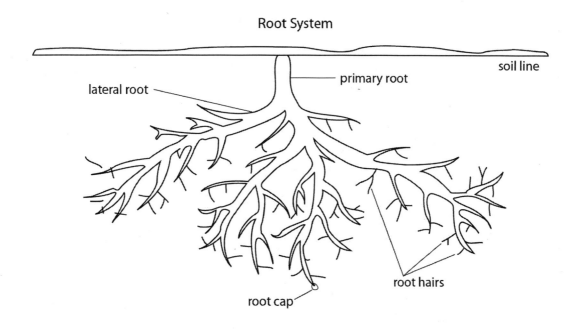

Root System

lateral root

primary root

soil line

root cap

root hairs

The Shoot System

The organs of the shoot system are the leaves, stem, buds, and flowers. Below, read about each organ in the shoot system and color each organ as instructed.

- **Leaves**—The job of this organ is to make food for the plant. *Leaves* are where photosynthesis occurs. <u>Color them green</u>.

- **Stem**—The *stem* has two jobs. It supports the plant. It is also important for ***translocation***. Translocation is the movement of food (glucose) from leaves, where it is made, to other parts in the plant where it is stored or used. In plants that have them, branches are a part of the stem. <u>Color the stem green</u>.

- **Buds**—Leaves, branches, and flowers grow from *buds*. <u>Color the buds yellow-green</u>.

- **Flowers**—*Flowers* are the reproductive organ of plants. <u>Color the flower pink</u>.

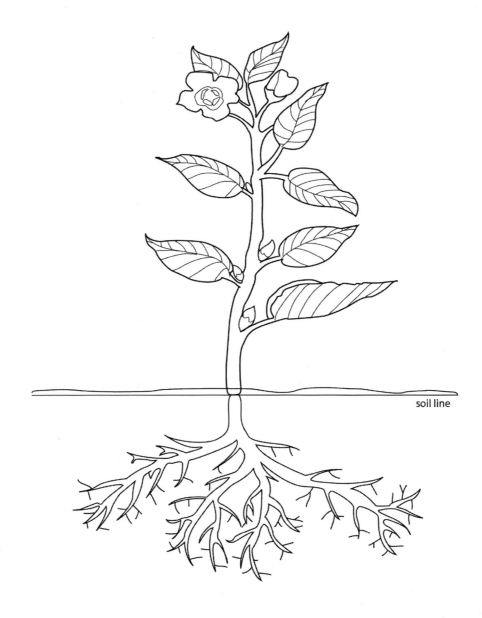

soil line

Plants: The Inside Story
Chapter 12: Dissection and Microscope Lab

Dissecting a plant is a lot less messy than dissecting a frog. You do not even have to wear gloves. Plants only have two organ systems, so there is less to look at too. You are going to perform an external examination of the anatomy. Then you are going to dissect both the root and the shoot systems.

Optional: You will make thin slices of some of the organs and look at them under the microscope.

Materials

- A plant with roots, shoots, leaves, and flowers
- X-Acto knife
- Running water to rinse dirt from roots

- Colored pencils
- Optional Microscope Supplies:
  - Microscope
  - Slides
  - Water
  - Slide cover
  - Syringe
  - Methylene blue

Procedure

1. If you are using a potted plant, pull it out of the soil. Be careful that you do not break the plant off at its root. Try to be VERY gentle so that some of the root hairs with root caps come with the primary and lateral roots.

2. Very gently, rinse any dirt away from the roots.

3. On My Plant Lab Sheet, draw your entire plant. You will be labeling the parts as you do the dissection.

4. Write the name of the type of plant on your lab sheet and label the root and the shoot systems.

Chapter 12: Dissection and Microscope Lab *continued*

Root:

5. Label the different parts of the root system on your plant drawing.

6. On the Dissection Lab Sheet, write your observations about the root system. As you follow the instructions (below) to cut the different parts, write your observations about each.

7. Carefully, with the X-Acto knife, cut a piece of thin, intact root (one that is not broken off at the end) about ½ inch long. The inner part of the root is called the vascular cambium. It has xylem and phloem running through it.

8. Microscope: Put the root piece on a slide. Place a drop of water and a slide cover on it, then look at it under the microscope. Scroll away from any brown clumps; the roots are whitish. Next, take off the slide cover and put a drop of methylene blue on the slide. Carefully rinse using water from the syringe and small pieces of paper towel to absorb the excess. Now you can look for root hairs. The larger root looks better without stain, but the root hairs show up better with stain. Next, look at the ends of your root. Do you think you were able to collect a root cap? If there is no root cap, the root looks like a torn rope. If there is a root cap, it looks like the end of the root has a webbed cap on it.

9. Cut another root off your plant. Choose a thick one. Rinse it again under water. Remember, roots are not brown; dirt is. Look at the place where you sliced the root. Can you see the rings of layers? You are looking at all three types of tissue in the root system. Dermal tissue surrounds the root. Ground tissue is below the dermal layer. Vascular tissue is in the center part of the circle; the vascular tissue transports materials to and from the roots.

10. Microscope: Make a thin slice of this circular piece. Stain it and look at it under the microscope. It will be thick. There is a very hard layer surrounding the vascular tissue that makes it difficult to get a thin slice. You might need to use top lighting with this slice.

11. Make a cut to separate the root system from the shoot system on your plant. Cut the main part of the root, the primary root, in half lengthwise and draw a cross section of it on your lab sheet. Now cut a smaller root. Cut it in half and peel off the outer layer. What part did you just peel off? Note the differences and similarities of the cross-sections.

Stem:

12. Microscope: Cut about a 2-inch-long piece at the juncture of the root and the shoot. Cut away the outer layers, down to the tender center. This is mostly vascular tissue. Examine it with your eyes. Then cut as thin a slice as you can, stain it, and look at it under your microscope.

13. Look at the cut end where the shoot system separated from the root system. Why is the shoot system green and not the root system?

Chapter 12: Dissection and Microscope Lab *continued*

14. Label the parts of the shoot system on your plant drawing. Write down your observations about the stem on the dissection sheet.

15. Microscope: Cut one of the younger stems from the plant. Cut it lengthwise. Make a thin slice, and make a wet mount slide. Look at it with the microscope. Notice how the sizes of the cells change depending on the layer.

Leaf:

16. Dissecting a leaf—the leaf is a good place to see all three types of tissue:

   • Dermal tissue (has stomata in it)

   • Vascular tissue (brings water to ground cells to be used in photosynthesis and takes glucose to the rest of the plant)

   • Ground tissue (has chloroplasts where photosynthesis takes place)

17. Cut off one of the leaves. Slice a thin slice from the cut end. Examine it with your eye or under a microscope. Take the X-Acto knife and CAREFULLY shave a small slice along the underside of the leaf, then peel it down as far as you can from the leaf. It will tear off at some point, which is what you want. You are peeling the dermal tissue away from the outside of the leaf. The dermal tissue is almost transparent.

18. Microscope: Make a wet mount slide with this layer. First, without stain and then with stain, see if you can find stomata in this layer. Cut lengthwise down the leaf. Take a thin slice along this cut. Do you see the thin outer layer of tissue? Why are there more chloroplasts around the top and bottom of this layer than in the center?

Bud:

19. Look for and label the buds on your drawing. Cut one of the buds off your plant. Remember the buds are the nodes out of which the shoot's leaves and flowers grow. It looks a lot like the stem from the inside, doesn't it? Record your observations on the dissection sheet.

Flower:

20. Pick one of the flowers, draw it, and record your observations. You will dissect a flower in the next chapter, so you do not need to cut it open.

Upcoming lab preparation: If you're using a dried bean in the Chapter 13 Lab, Flower and Seed: Inside View, it needs to be soaked in room-temperature water the night before starting the lab.

Plants: The Inside Story

Chapter 12: Dissection and Microscope Lab Sheet

Name_____ **Date**_____

My Plant

Type of Plant: _____

Chapter 12: Dissection and Microscope Lab Sheet *continued*

Plant Dissection

Record your observations below, using words and drawings.

Root:

Leaf:

Chapter 12: Dissection and Microscope Lab Sheet *continued*

Stem:

Bud:

Flower:

Pandia PRESS

Isabella Abbott

Chapter 12: Famous Science Series

Isabella Abbott

Famous Botanist

What is Isabella Abbott's given name? What does her given name mean? When and where was she born?

She was the first native Hawaiian woman to receive what college degree?

In 1982, Abbott was hired by the University of Hawaii to study ethnobotany. What does the term *ethnobotany* mean?

Abbott was one of the world's leading experts on what type of organism? Many people think this organism is a plant. It isn't. What type of organism is it? Why isn't it a plant?

Plant Anatomy

Chapter 12: Show What You Know

1. Match the vocabulary word on the left side with the definition on the right side.

dermal tissue ◯	◯ primary root, lateral root, root hairs
vascular tissue ◯	◯ leaves, buds, stems, flowers
ground tissue ◯	◯ covers and protects the outside of plants
root system ◯	◯ tiny cells where water is absorbed from the soil
shoot system ◯	◯ shoots, leaves, and flowers grow from them
primary root ◯	◯ protects the root tip as it grows through the ground
lateral roots ◯	◯ provide support and allows for the flow of water and nutrients
leaves ◯	◯ reproductive organ in plants
root hairs ◯	◯ organ where photosynthesis takes place
root cap ◯	◯ large root with smaller ones growing out of it
buds ◯	◯ transports food, water, and nutrients throughout the plant
stems ◯	◯ small roots growing out of primary root
flower ◯	◯ what most of the plant is made from

Chapter 12: Show What You Know *continued*

2. Label the parts of the plant (listed below) on the drawing. Color the parts, if you like.

<u>Root System</u>

○ Primary root
○ Lateral root
○ Root hairs
○ Root cap

<u>Shoot System</u>

○ Stem
○ Bud
○ Flower
○ Leaf

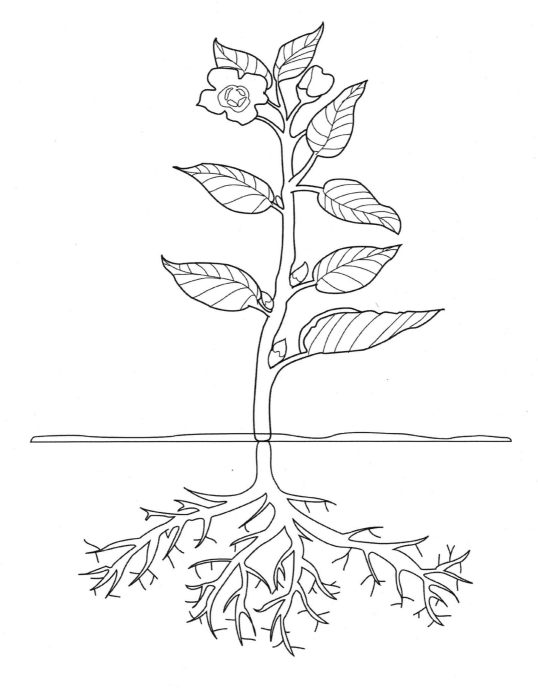

Chapter 13: Plant Reproduction

Explore

Flower and Seed: Inside View
Chapter 13: Dissection Lab

The day before: If your bean is not fresh (i.e., it's a dried bean), it needs to be soaked in room temperature water overnight.

Angiosperms have flowers and produce seeds. You have probably admired many different flowers in your life. And you have most likely eaten many seeds. Beans are seeds, and most berries cannot be eaten without eating their seeds. In all that time, have you ever wondered what the flower you were admiring or the seeds you were eating looked like inside? Today you will dissect a flower and a seed and do just that. When choosing a flower to dissect, choose a perfect flower. Perfect flowers have male and female parts together in the same flower. A big, showy flower with a large pistil and stamens is best. You also need a big seed. A large lima bean is a good choice to look at for a seed.

Materials

- Large perfect flower, with the male and female parts (lilies are great). Flowers to avoid are daisies, asters, calla lilies, roses, and irises.

- Lima bean
- X-Acto knife
- Scissors
- Tape

- Magnifying glass

Procedure

1. If the bean is not fresh, it needs to be soaked in room temperature water overnight.

2. As you read about each part, tape and label the dissected part to the lab sheet. For example, when you separate the petals and stamens, tape one of each to the lab sheet.

Chapter 13: Dissection Lab *continued*

The Angiosperm

3. Carefully remove the stamens and the petals. Look at the petal with the magnifying glass, then tape it to your lab sheet.

4. Cut off the stamens. Tape a stamen to the lab sheet. On your lab sheet, shake pollen from a stamen and put a piece of tape over it.

5. Examine a little bit of pollen with a magnifying glass. Draw a picture of a magnified pollen grain.

6. Cut away all parts, except the pistil, from the stem. Cut the pistil from the stem below the ovary. Examine the stigma with the magnifying glass.

7. Cut the pistil in half from the top, the stigma, down through the center of the style through the ovary base. This gives you a good view down the inside of the flower.

8. With a magnifying glass, look at the style. Look for ovules in the ovary. With the tip of the X-Acto knife, dig out some ovules and look at them with the magnifying glass. Draw the magnified ovules. Tape the piece of pistil you cut away to the lab sheet.

The Seed

9. Take the bean out of water. Carefully, using the X-Acto knife, split the bean into its two halves at the seam on the thin side of the bean. Compare your seed to the illustration of the dicot seed found on the lab sheet. Identify each of the parts. Look carefully at each part with the magnifying glass. Write a description of your seed below each part label.

The world's largest seed is a double coconut. A single seed can be nearly 3 feet in circumference and weigh 40 pounds.

Pandia PRESS

Flower and Seed: Inside View

Chapter 13: Dissection Lab Sheet

Name_____ Date_____

My Flower's Parts

Sepals

Petal

Stem

Female Part

Stigma

Style

Pistil (carpel)

Ovary

Female gametes are ovules

Male Part

Pollen grains are male gametes

Anther

Stamen

Filament

Chapter 13: Dissection Lab Sheet *continued*

My Seed's Parts

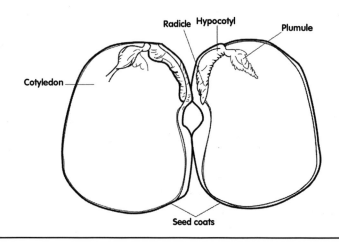

Cotyledon

Radicle Hypocotyl

Plumule

Seed coats

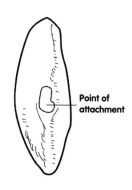

Point of attachment

Pandia PRESS

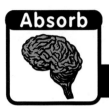 **Absorb**

Sunflower

Chapter 13: Famous Science Series

Famous Flower

If you drove a car through Russia, Ukraine, Hungary, and Austria, you would see miles and miles of fields planted with sunflowers.

Where did sunflowers originate?

Sunflowers have been found at archaeological sites. How old are the remains?

Today, what country is the number one consumer of sunflowers?

Sunflowers became important in this country because of two religious holidays. Name the country, the holidays, and why the sunflower is important to this country.

Chapter 13: Famous Science Series *continued*

The rulers of this country sent soldiers into battle with packages of sunflower seeds in what quantity?

In 1986, workers at the Chernobyl nuclear power plant caused an explosion that released massive amounts of radioactive material into the surrounding environment. How were sunflowers used to help clean up this problem?

Pandia PRESS

Plant Reproduction

Chapter 13: Show What You Know

1. Match the vocabulary word on the left side with its definition on the right side.

pistil ◯

◯ flowering plant

seed ◯

◯ top part of the pistil where pollen is trapped

petals ◯

◯ reproductive organ of angiosperms

stamen ◯

◯ female reproductive structure of an angiosperm

anther ◯

◯ where the ovules are stored in an angiosperm

pollen ◯

◯ used by mosses and ferns in reproduction

ovules ◯

◯ male reproductive structure of an angiosperm

flower ◯

◯ where pollen is produced for an angiosperm

angiosperm ◯

◯ male gametes

gymnosperm ◯

◯ survival capsules, seed coat on outside and food inside

cones ◯

◯ part of the pistil that joins the stigma and the ovary

style ◯

◯ the part of the stamen that supports the anther

fruit ◯

◯ when a seed first begins to grow

ovary ◯

◯ the part of an angiosperm that attracts pollinators

filament ◯

◯ protects seeds and aids in their dispersal

spore ◯

◯ female gametes

stigma ◯

◯ seed-producing plants that do not produce fruit

pollen tube ◯

◯ reproductive organs of gymnosperms

germination ◯

◯ tube that grows through the stigma to the ovary when pollen comes in contact with the stigma

Chapter 13: Show What You Know *continued*

2. Label the flower.

Parts to label:

○ Stem

○ Sepals

○ Petal

○ Stigma

○ Style

○ Ovary

○ Ovules

○ Filament

○ Anther

○ Pistil

○ Stamen

○ Pollen grains (draw)

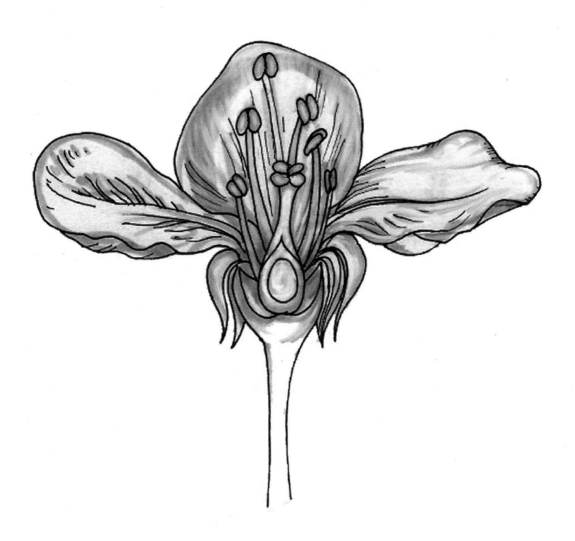

Pandia PRESS

Chapter 14: Nervous and Senory Systems

Read

I'll Do the Thinking 'Round Here

Chapter 14: Lesson 1 Activity

The following coloring activity is found in Chapter 14 Lesson 1 in your Textbook.

Label and Color the Nervous System

As you read about the organs in the nervous system, label the different pasts and color them on the illustration below.

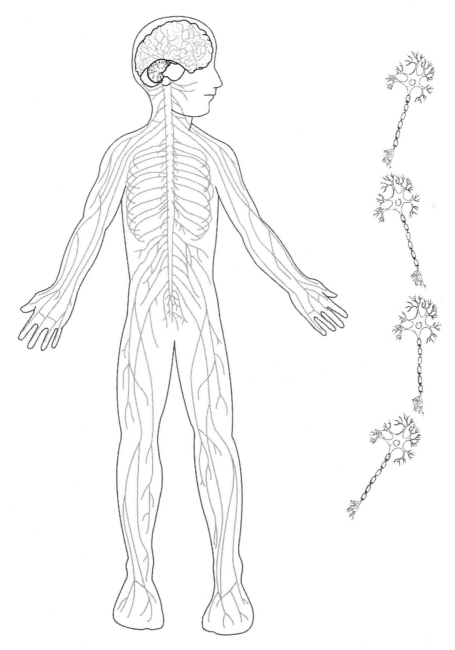

Absorb

Count Alessandro Volta

Chapter 14: Famous Science Series 1

Allesandro Volta

Famous Frog Legs

In 1771, biologist Luigi Galvani was in his lab cutting a frog's leg with a metal scalpel. What happened when he did?

What did he think caused it to happen?

Alessandro Volta was a friend and fellow scientist. When he heard of Galvani's results, he performed the experiment. He got the same results. However, Volta did not agree with Galvani's conclusions. What did Volta think caused the frog's leg to twitch?

Volta continued with his experiments. His experiments led him to invent something in 1800 that we still use to this day. What is it?

Does your body produce electrical signals? Explain your answer.

Pandia PRESS

Explore Your Brain Is Not a Battery

Chapter 14: Lab 1

Alessandro Volta was so excited when he learned of the results of Galvani's experiment that he decided to repeat the experiment. When he did, he realized Galvani was wrong about how he interpreted his results. In science, it is important that different experimenters be able to perform the same experiment and get the same results with the same interpretations. When this does not happen, scientists work on the problem until they figure out who is correct. In this case, Galvani was sure he was right, so he did not keep looking for a better answer. Volta kept looking, and in the process invented the battery.

Your brain is not generating electrical currents to make you respond, as Galvani believed. Your body sends electrical and chemical signals using water and molecules dissolved in the water.

If you cut open a lemon, it is wet from water and dissolved molecules. Is that all it takes to make a battery? A lemon does not have a nervous system, like a person or a frog does. Can a lemon be used to make a battery? Today you will find out the answer to that question. When you turn on a calculator you turn a calculator on, the batteries generate an electrical charge that turns on the calculator. If a lemon can be used to make a battery, it should be able to turn on a calculator too.

CAUTION: *The lemon battery you will be making is a weak battery. Do not try connecting the wires to a real (store-bought) battery. This can be damaging to equipment and hazardous to you.*

Materials

- 1 juicy lemon
- 3 pieces of wire, each 8 to 12 inches long
- 2 zinc nails, screws, or galvanized nails
- 2 pennies from 1983 or before, OR 2 pieces of copper wire rolled up like paper clips
- Calculator that requires one battery
- Salt
- Knife
- Paper towels to wipe up spilled juice

Chapter 14: Lab 1 *continued*

Procedure

1. Cut the lemon in half. Cut a small slit in each lemon half, and insert the copper wire or pennies into it, one in each lemon. Keep your area clean, wiping up spilled juice.

2. Strip the ends of the wire to about 3 in. Wrap the end of one piece of wire around one piece of copper. Do the same with the other. Insert the nail/screw, one into each lemon. Do this at the other end of the lemon from the copper. The copper and zinc cannot touch or the experiment will not work. Connect one of the lemons to the other by connecting the wire wrapped around a copper piece to the nail/screw in the other lemon.

3. Sprinkle salt on the top of each lemon. This increases the amount of dissolved molecules to send electrical signals.

4. Connect the remaining unattached wire to the nail/screw with no wire wrapped around it.

5. Take the back battery compartment cover off the calculator. Take the battery out of this compartment. Identify the positive and negative ends of the compartment. If the battery is a small circular battery (calculators that use this kind of battery are the best for this experiment) look at which end is up when the battery is placed correctly in the battery compartment. You might need to look for a plus sign on the battery to determine this.

6. There are two wires, one from each lemon, that are only attached on one end of the wire. In the back of the calculator, where the battery was taken out, connect the wire that is wrapped around the piece of copper to the positive side of the battery compartment. Connect or touch (it helps to have another person do this, so you can operate the calculator) the wire from the zinc to the negative side of the battery compartment. Do not let these two wires touch each other. The calculator is ready for use. Answer the questions on the lab sheet.

Troubleshooting

Your lemon is a very weak battery. It does not have enough power to light a light. If you want to light a flashlight bulb or even a Christmas tree light, you need at least four lemon pieces connected together. If your calculator isn't turning on, try these tips:

- Make sure you have correctly identified the positive and negative ends of the calculator.

- Spilled lemon juice or wires crossing over and touching each other can keep the battery from working.

Your Brain Is Not a Battery

Chapter 14: Lab 1 Sheet

Name_____ **Date**_____

Electricity is energy produced by a flow of electrons. Chemical reactions that produce electrons occur in batteries. Batteries have a positive end and a negative end. Electrons are negatively charged so they flow from the negative end to the positive end of a battery. It is a case where opposites attract. Electrons flow through materials that are conductors. Copper wire and water are both great conductors. When you connect wire from one end of the lemon battery to the other end, electrons flow from the negative end to the positive end through the wire, and a circuit is created. If you attach a calculator in the middle of the circuit, it will be powered by the flow of electrons, by the electricity created.

You are 65 percent water. Use the above discussion and the nervous system illustration you colored earlier to describe how signals are sent to and from your brain.

How would dehydration affect your brain's ability to respond to signals? Or, what would happen if the width of the wire in your lemon battery is increased or decreased?

Come to Your Senses

Chapter 14: Lesson 2 Activity

The following coloring activity is found in Chapter 14 Lesson 2 in your Textbook.

Color the Sensory Systems

As you read about each of the four senses, color and mark the sensory system illustrations below.

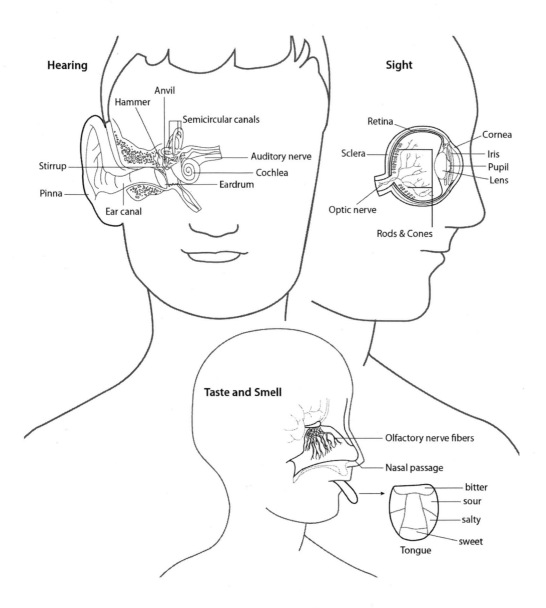

Explore

Seeing Sound Waves

Chapter 14: Lab 2

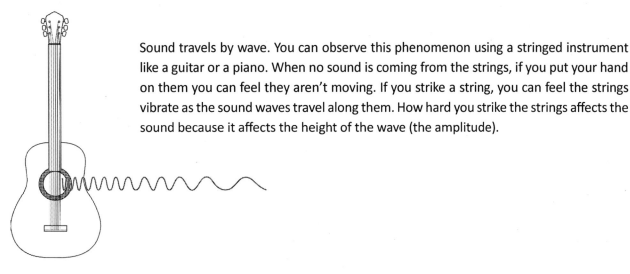

Sound travels by wave. You can observe this phenomenon using a stringed instrument like a guitar or a piano. When no sound is coming from the strings, if you put your hand on them you can feel they aren't moving. If you strike a string, you can feel the strings vibrate as the sound waves travel along them. How hard you strike the strings affects the sound because it affects the height of the wave (the amplitude).

Materials

- Rubber band small enough to tightly span between your two fingers

- Stringed instrument (examples: guitar or piano)

Procedure

1. Put the rubber band spanning around your thumb and index finger, the tighter the better. Pluck the rubber band, keeping your fingers still and the distance constant. Touch the rubber band with the back of one of the fingers from your other hand. Write your observations down on your lab sheet.

2. Put your middle finger and your index finger alongside of your throat. Make a variety of sounds: humming, coughing, saying "ah" repeatedly, and talking. Write your observations down on your lab sheet.

3. Touch the strings of the instrument when no sound is coming from it. Then strike the instrument, keys or string, and put your hand on the strings. Strike the guitar harder and softer, varying the sound. Write your observations down on your lab sheet.

Seeing Sound Waves

Chapter 14: Lab 2 Sheet

Name_____ Date_____

1. Describe what you saw, heard, and felt when you struck the rubber band.

2. Using what you learned in the lemon battery lab, describe how sound waves become electrical signals your brain can interpret.

3. What did you feel when you touched the side of your throat? In words or with a picture, describe how spoken words are heard. Start with the source (the throat) and end with the sound going into an ear to the brain.

4. What is the process by which sound gets from its source to being heard by you?

Pandia PRESS

Chapter 14: Lab 2 Sheet *continued*

5. Describe what you saw, heard, and felt when you struck the instrument.

6. When you put your hand on the strings the sound started to quiet. Why?

7. How do you think bats "see" in the dark? Hint: It is called *echolocation*. Can you describe what happens?

Pandia PRESS

Absorb

Cochlear Implants

Chapter 14: Famous Science Series 2

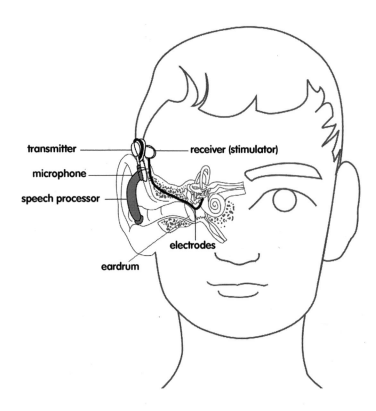

The first person to realize that electrical stimulation of the auditory system could simulate sound was Alessandro Volta in 1790. He put metal rods in his own ears and sent an electrical charge through them. He heard a noise like "thick boiling soup." Are you thinking what I am: Where was his mother when he was doing this experiment?!

Chapter 14: Famous Science Series 2 *continued*

Describe how sound travels from its source to you so you can hear the sound.

What happens if the cochlea in the ear does not work?

A cochlear implant is surgically implanted under the skin behind the ear. How does it work?

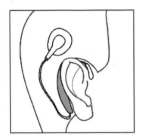

How old do you have to be in the United States before you can get a cochlear implant?

Nervous and Sensory Systems
Chapter 14: Show What You Know

1. Fill in the Blanks. What are the three main parts of your brain?

1.

2.

3.

The _____ controls involuntary vital functions.

The_____ is responsible for complex thought.

The _____ coordinates movement.

2. Match the word with its definition.

Cornea ◯   ◯ has tiny muscles that control size of eye

Pupil ◯   ◯ white part of eye

Iris ◯   ◯ clear layer outside of eye

Lens ◯   ◯ behind pupil, focuses light to back of eye

Retina ◯   ◯ light receptors

Optic Nerve ◯   ◯ where images are formed

Sclera ◯   ◯ black hole in the center of eye

Rods and Cones ◯ ◯ nerve going from eye to brain

The above make up your _____, which is your organ of _____.

 Pandia PRESS

Chapter 14: Show What You Know *continued*

3. Label the parts of the ear using the list below.

hammer/malleus

pinna

semicircular canals

auditory nerve

stirrup/stapes

cochlea

eardrum

ear canal

anvil/incus

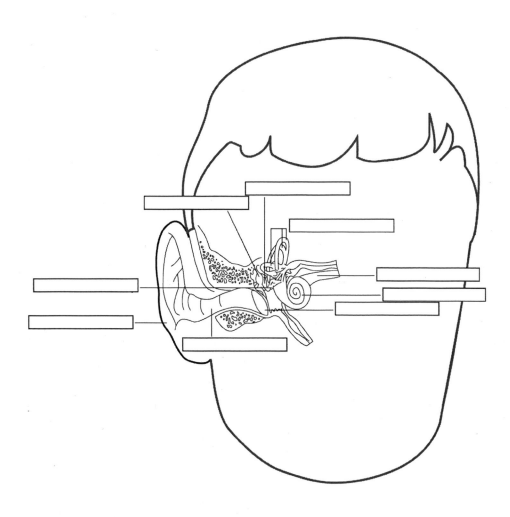

Chapter 14: Show What You Know *continued*

4. Multiple Choice

1. The parts of your inner ear are:

   ○ cornea, iris, pupil
   ○ malleus, incus, stapes
   ○ semicircular canals, cochlea, auditory nerve
   ○ pinna, ear canal, eardrum

2. When signals travel through your nerves, the path they take is:

   ○ dendrite, axon, main body part
   ○ axon, main body part, dendrite
   ○ dendrite, main body part, axon
   ○ main body part, axon, dendrite

3. Scent molecules enter your nose where there are _____ that send information about them to your brain.

   ○ auditory nerves
   ○ olfactory nerves
   ○ optic nerves
   ○ nasal nerves

4. The organs of your nervous system are:

   ○ brain, spinal cord, nerve cells
   ○ brain, heart, veins
   ○ eye, ear, nose, tongue
   ○ axon, main body part, dendrites

5. Sound and light travel through the air as

   ○ waves.
   ○ pixilated molecules.
   ○ electrical signals.
   ○ chemical reactions.

6. Chemical receptors on your tongue determine if something is

   ○ bad for you.
   ○ food your body needs to build cells.
   ○ healthy.
   ○ sweet, sour, bitter, and/or salty.

7. The parts of your middle ear are:

   ○ dendrite, axon, main body part
   ○ malleus, incus, stapes
   ○ semicircular canals, cochlea, auditory nerve
   ○ pinna, ear canal, eardrum

8. Your body's main nerve is the

   ○ axon.
   ○ main body part.
   ○ spinal cord.
   ○ dendrite.

9. Nerve cells send messages through your body using

   ○ waves.
   ○ pixilated molecules.
   ○ electrical signals.
   ○ ether.

10. Four organs of sense are:

   ○ brain, heart, lungs, nerves
   ○ tongue, teeth, nose, mouth
   ○ ear, eye, nose, tongue
   ○ brain, axon, dendrite, spinal cord

Pandia PRESS

Read

Your Skin and What's Within
Chapter 15: Lesson 1 Activity

The following coloring activity is found in Chapter 15 Lesson 1 in your Textbook.

Color the Integumentary System. *As you read about each organ, color and label the illustrations below.*

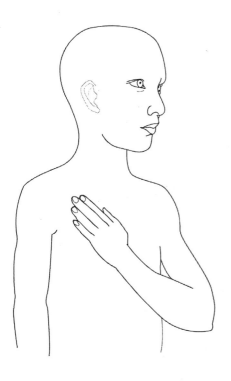

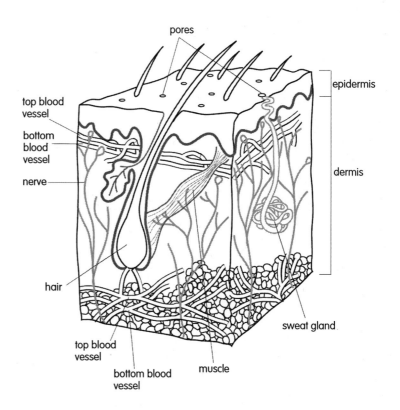

Explore
Reading with Your Fingers
Chapter 15: Lab 1

*You will need someone else to do the prep work for this lab. This lab will only work if you do not know what is written.

You read with your eyes, your sense of sight. Using Braille, people who are visually impaired read with their sense of touch. You are both using one of the senses hardwired into your nervous system to interpret symbols and put them together and make a meaning from them. Your nervous system sure is smart, isn't it? You will not be using the Braille alphabet in this lab. You will use the Standard English alphabet. Do you think you will be able to read with your fingers?

Materials

- Piece of corrugated cardboard
- Nail
- Blindfold

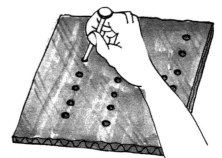

Poke the nail through the back of the cardboard so the front side displays the word.

Procedure

1. This part needs to be done by someone other than the reader. Using print letters, write a word or words on the cardboard. Take the nail and, from behind, punch raised circles in the shape of the letters. If you do not do it from behind, the letters will not face the correct direction. Do not punch all the way through the cardboard. Make the space between the raised punches about as wide as the raised punches are.

2. Do not tell the reader what the word is. Blindfold the reader. Put the cardboard face up so the punches are raised and the letters face the correct direction. Let the reader see if they can read with their sense of touch. There is no lab sheet for this lab.

Pandia PRESS

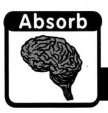

Absorb

Louis Braille

Chapter 15: Famous Science Series 1

Famous Alphabet Inventor, Louis Braille

When and where was Louis Braille born?

How did Braille lose his eyesight?

At the age of ten, Braille was awarded a scholarship to where?

What two instruments did Braille learn to play at school?

Braille invented an alphabet. Who uses it and how does it work?

Is Braille still in use today?

Was the Braille alphabet accepted when it was invented?

Explore

The Skinny on Skin
Chapter 15: Microscope and Lab Sheet

Name_____ **Date**_____

Specimen _____

Type of mount_____ Type of stain used_____

Materials

- Microscope
- Slide
- Slide cover
- Water

- Your arm or leg (whichever has the driest skin)
- Methylene blue
- Tissue
- Butter knife

Procedure

1. The slide needs to be VERY clean. Look at it under the microscope before scraping skin onto it.

2. Use the butter knife to gently scrape the top side of your arm or leg. Wipe the edge with scrapings on it onto the slide. Look for some flakes on the slide. You are collecting your own skin cells. Put a drop or two of water on the slide. Put a drop of methylene blue on the slide.

3. Use the tissue to soak up the water and methylene blue. Put a drop more water on the slide.

4. Now hunt for a good cell. Maybe you can find a hair with the follicle still attached.

5. Draw your favorite view at 400x magnification below.

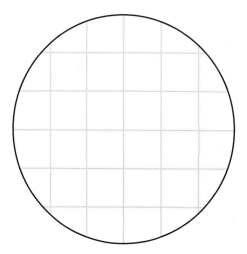

Building Cells and Ridding Waste
Chapter 15: Lesson 2 Activity

The following coloring activity is found in Chapter 15 Lesson 2 in your Textbook.

Color the Digestive System and Urinary System

As you read about the digestive and urinary systems, color and mark the illustrations below.

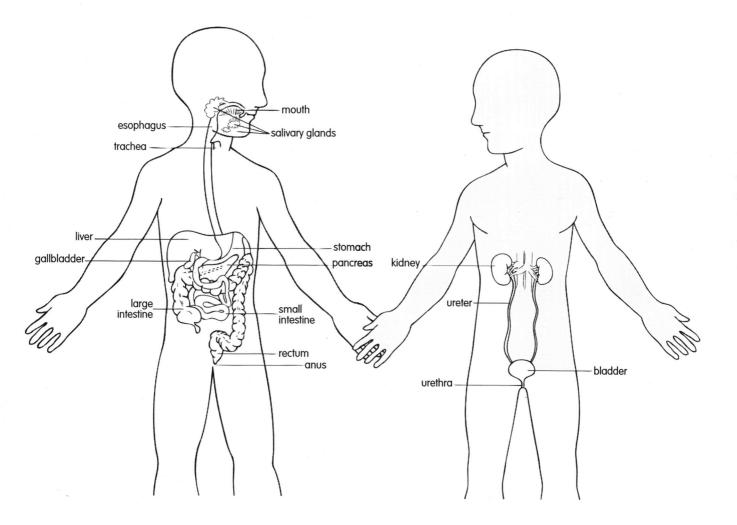

Explore

Kidneys Clean Up

Chapter 15: Lab 2

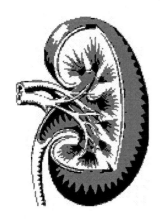

Did you know that you must have at least one functioning kidney to live? Your body has to be able to filter the waste from your blood. In fact, your kidneys filter your blood about 400 times a day! Your kidneys have about 2 million tiny filters, called **nephrons**. Waste from the nephrons combines with water to make a solution, called urine, that flushes waste from your body.

Do you know how filters work? Do you think you can filter something as well as your nephrons? Can you filter all the waste from dirty water? I bet your nephrons could. Today you will set up a filtration system and look at how filters work.

Materials

- 2 liter plastic soda bottle
- Sink
- Scissors
- 5 coffee filters
- 1 to 2 cups of clean gravel
- 1 to 2 cups of clean sand
- 15 to 20 cotton balls

- Dirty water in a bowl, just a few suggestions for making it dirty: dark cooking oil (e.g. olive), pieces of paper, dirt, and/or wood chips. Play around with this.
- Camera or colored pencils

- Optional: other filtering material that you can think of and find around your house

Pandia PRESS

Chapter 15: Lab 2 *continued*

Procedure

1. Cut the bottle about one-third of the way down from the top. Put the top one-third of the bottle upside down into the bottom, like a funnel.

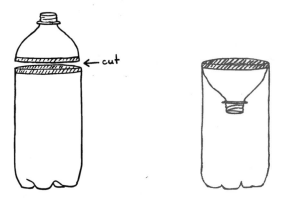

2. Make the dirty water in a bowl. Write the ingredients in the water on your lab sheet.

3. Now, you are going to design your filter by layering materials in the top upside part of the bottle. Your filter materials are coffee filters, gravel, sand, and cotton balls. Each different material will be layered in the filter. You can use all the filter materials, or three, or two, or only one. Choose the material you are going to use and where you want each material layered in the filter. Think about what is in your water and what you are trying to filter out of it. Use the amount, type, and the order of filter material that you think will do the job without going overboard. If it doesn't filter the water clean the first time, redo the experiment until it does, adding more filter material or changing the order until you find the perfect amount and combination.

4. Slowly (don't just dump it in), pour the dirty water through the filter.

5. How clean is your water after filtering? Write your observations on your lab sheet.

6. When you have created a filter that works (i.e., your water comes out mostly clean), draw or take a picture of your setup for your lab sheet. Label each layer, even if you can't see the layer. Briefly, write your reasoning for the location, the amounts, and the inclusion of each layer.

7. Take your filter out of the bottle carefully. Try to identify which filter material took out the different components of the dirty water. Complete the lab sheet.

8. Optional: Make other filters with clean material and dirty water. If you make another filter, try to make one that improves on the first one. Ask yourself: Which filter material seems to be an important filtering agent, and which does not seem to add much to the filter?

9. Optional: If you are having trouble getting the water clean, set up a series of filters using more bottles, so that you filter out material in successive steps.

Kidneys Clean Up

Chapter 15: Lab 2 Sheet

Name_____ Date_____

Materials Used:

Procedure – How did you layer the filter materials? Why did you use that order?

Illustration:

Observations and Results:

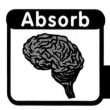

Willem Johan Kolff

Chapter 15: Famous Science Series 2

The most famous doctor you have never heard of—Willem Johan Kolff

Willem Johan Kolff

When and where was Willem Kolff born?

Kolff was a doctor when the Nazis invaded the Netherlands. What did Kolff do when they invaded?

Kolff set up a special kind of bank. What was it?

How did Kolff save people from the Nazi labor camps?

Chapter 15: Famous Science Series 2 *continued*

In 1938, Kolff watched a young man die a painful death because his kidneys were failing. If your kidneys stop working, they do not filter waste from your body. The waste poisons your blood. Kolff and other doctors realized if they could filter the poisons out of the blood, they could save people from these painful deaths. What did Kolff do to address this problem?

Next, Kolff invented a membrane-oxygenator that is still used today. What does it do? What surgery is it still used in?

Kolff and one of his students, Dr. Robert Jarvik, invented another lifesaving device called the Jarvik 7. What is it and what did it do?

Integumentary, Digestive, and Urinary Systems

Chapter 15: Show What You Know

Multiple Choice

1. Your largest organ is your

   ○ liver.

   ○ brain.

   ○ heart.

   ○ skin.

2. Which part of your brain controls your digestive and urinary system?

   ○ Cerebrum

   ○ Cerebellum

   ○ Medulla

   ○ Spinal cord

3. Which nerves connect your skin to your brain?

   ○ Auditory nerves

   ○ Peripheral nerves

   ○ Optic Nerves

   ○ Olfactory nerves

4. Gastric juices in your stomach

   ○ come from food.

   ○ are dangerous.

   ○ are flushed into your pancreas.

   ○ break up food and kill bacteria.

5. Your urinary system

   ○ filters waste from your body that your cells don't use.

   ○ carries water to your cells.

   ○ digests food.

   ○ extracts vitamins from food.

6. Skin helps maintain homeostasis by

   ○ regulating hydration.

   ○ maintaining body temperature.

   ○ cleaning your pores.

   ○ helping hair grow.

   ○ All of the above

   ○ Answers 1 and 2

   ○ Answers 2 and 3

7. Your blood carries molecules to your _____, where they are sent to cells.

   ○ kidneys

   ○ esophagus

   ○ large intestine

   ○ liver

8. Melanin in the epidermis

   ○ protects skin from sunlight.

   ○ determines skin color.

   ○ causes freckles.

   ○ All of the above

Chapter 15: Show What You Know *continued*

9. The organ that filters your blood:

   ○ kidney
   ○ gallbladder
   ○ liver
   ○ ureter

10. Food is broken down to small molecules your body can absorb in your

   ○ liver.
   ○ small intestine.
   ○ gallbladder.
   ○ bladder.

11. Your body makes saliva

   ○ so you can spit.
   ○ so you don't get infections.
   ○ to help you swallow.
   ○ to help break down carbohydrates.

12. This organ is lined with muscles that mash up food into smaller bits:

   ○ gallbladder
   ○ stomach
   ○ large intestine
   ○ esophagus

13. The color of your urine and poop comes from

   ○ digested red blood cells.
   ○ the food you eat.
   ○ chemicals in your body.
   ○ bacteria.

14. This is the last stop before your rectum. This is the organ that extracts the last bit of vitamins and water from your food:

   ○ pancreas
   ○ kidneys
   ○ large intestine
   ○ stomach

15. The organ that fills up and then tells your brain you need to pee is the

   ○ ureters.
   ○ urethra.
   ○ bladder.
   ○ kidneys.

16. Your skin has two main layers. They are

   ○ the dermis and muscle.
   ○ the dermis and subcutaneous fat.
   ○ the epidermis and dermis.
   ○ the nerves and epidermis.

Questions

The three main organs of the integumentary system are:

1.

2.

3.

Your immune system is the system that protects you from getting infections. Your skin is sometimes called the first line of defense for your immune system. What is meant by this?

Pandia PRESS

Chapter 16: Endocrine and Reproductive Systems

Read

The Balancing Act

Chapter 16: Lesson 1 Activity

The following coloring activity is found in Chapter 16 Lesson 1 in your Textbook.

Color the Endocrine System. *As you read about the endocrine system, color the illustration below.*

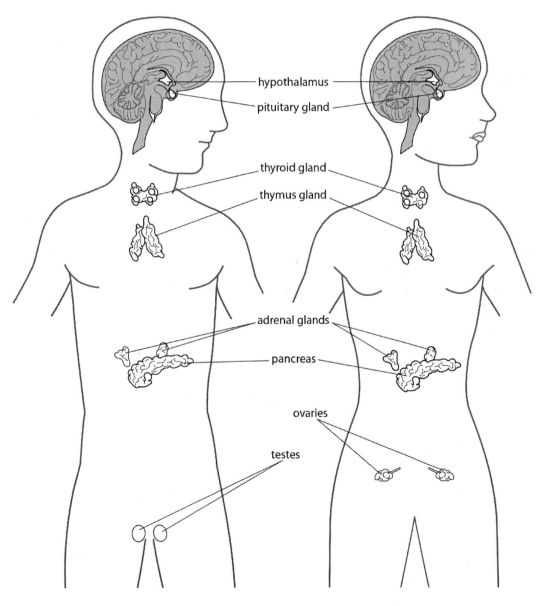

hypothalamus

pituitary gland

thyroid gland

thymus gland

adrenal glands

pancreas

ovaries

testes

Male Endocrine System   Female Endocrine System

Pandia PRESS

Explore
Let's Maintain Homeostasis
Chapter 16: Lab 1

It is noon. I am so hungry for lunch.

The endocrine system maintains homeostasis in humans. It regulates internal conditions in your body to keep them within the normal range.

Your endocrine system has been maintaining homeostasis all these years without you even thinking about it. You just do whatever it tells you to do, like eat and drink. You are so obedient! Your endocrine system maintains homeostasis using *homeostatic feedback mechanisms*. Feedback mechanisms work like a thermostat in your home. You set the thermostat at a certain temperature. If the temperature goes higher than the set temperature, the air conditioner turns on until the temperature lowers to the set temperature. If the temperature goes lower than the set temperature, the heater turns on until the temperature rises to the set temperature.

The mechanism your body uses to control the amount of glucose in your blood is an example of a homeostatic feedback mechanism. Your cells need glucose for energy. It is important that your blood has enough glucose to meet your energy needs, but not too much. If you eat a meal high in carbohydrates, glucose levels in your blood increase. When this happens, your pancreas makes insulin. Insulin causes a decrease of glucose in your blood. If the glucose level in your blood gets too low, your pancreas makes glucagon. Glucagon causes an increase of glucose in your blood. This brings the glucose level in your blood back to normal.

Today you are going to pay more attention to your endocrine system and homeostasis. You will see your endocrine system in action. To maintain homeostasis, your body monitors what is occurring inside and outside of it, keeping internal conditions within their normal range. When conditions approach the high or low end of this range, your endocrine system sends feedback to keep conditions within the normal range.

Pandia PRESS

Chapter 16: Lab 1 *continued*

Materials

- Room with a thermostat
- Flashlight

- Another person (a test subject)
- Watch with a second hand or a timer

Procedure

1. Turn the thermostat in your house or classroom down or up, just until it turns on. Wait for it to turn off again. Complete #1 on the lab sheet.

Now that you know how it works for a thermostat, let's look at how homeostasis works for a person.

2. Look at the eye pupils of your test subject. Shine the flashlight very close to, but not directly in, the eyes of your test subject. Complete #2 on the lab sheet.

3. Now let's test your endocrine system. Balance on one foot. Do not hold on to anything. Complete #3 on the lab sheet.

4. Take your pulse rate by putting your index and middle fingers on your wrist. DO NOT use your thumb; it has its own pulse. Put your finger toward the thumb side of your wrist until you feel your pulse. You will need to apply some pressure by pushing down slightly. Count the number of beats in ten seconds. Multiply this number by 6 to determine the beats per minute, and record this on your lab sheet. Run in place, run around the room, or do jumping jacks for one minute. At the end of this time, determine your pulse rate again and record it on your lab sheet. Lie on the floor or sit somewhere comfortable and completely relax for five minutes. Determine your pulse rate and record it on your lab sheet. Finish the questions for #4 on your lab sheet.

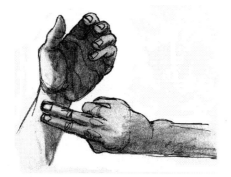

Let's Maintain Homeostasis

Chapter 16: Lab Sheet

Name_____ Date_____

1. Starting Temperature:_____   Ending Temperature:_____

   How does this demonstrate a feedback mechanism?

2. What happened to the pupils of the person when you shined the light in their eyes? Describe the pupil before and after in words and illustrations.

   Before   After

   Explain how that is an example of maintaining homeostasis.

3. What happened when you balanced on one foot? What happened when you swayed? How is that an example of maintaining homeostasis?

Pandia PRESS

Chapter 16: Lab Sheet *continued*

4. Beginning Pulse Rate: Pulse Rate after being active: Ending Pulse Rate:

_____ x 6 = _____   _____ x 6 = _____   _____ x 6 = _____

What happened to your pulse rate when you were moving quickly? What happened to your pulse rate while you rested?

Your pulse rate measures how fast your blood is pulsing through your body. When you are active, the cells in your body need more oxygen, which your blood carries to them. How does your pulse rate increasing when you are active and decreasing when you are at rest demonstrate a feedback mechanism?

Body Weight and Homeostasis
Chapter 16: Famous Science Series 1

One of the most important jobs your endocrine system has is maintaining homeostasis. What things does your endocrine system monitor for homeostasis?

Whatever weight I am at, that's the weight I want to stay.

When a person has completed puberty and is finished growing, their body tries to maintain its weight. Why is staying at a weight part of homeostasis?

What is the job of the hypothalamus?

How can it become a problem that maintaining your weight is part of homeostasis?

To Be

Chapter 16: Lesson 2 Activity

The following coloring activity is found in Chapter 16 Lesson 2 in your Textbook.

Color the Reproductive Systems and Zygote to Birth. *As you read about reproduction, color the following illustrations.*

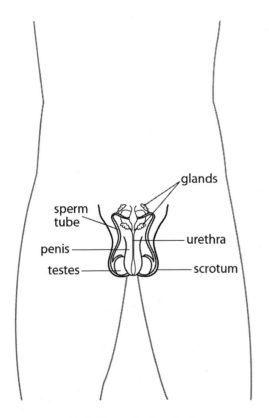

Male Reproduction System

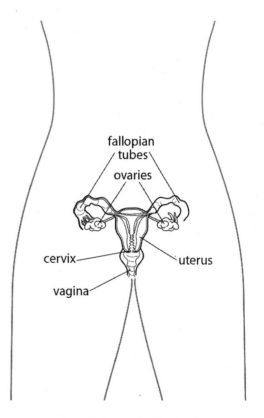

Female Reproduction System

Zygote to Birth

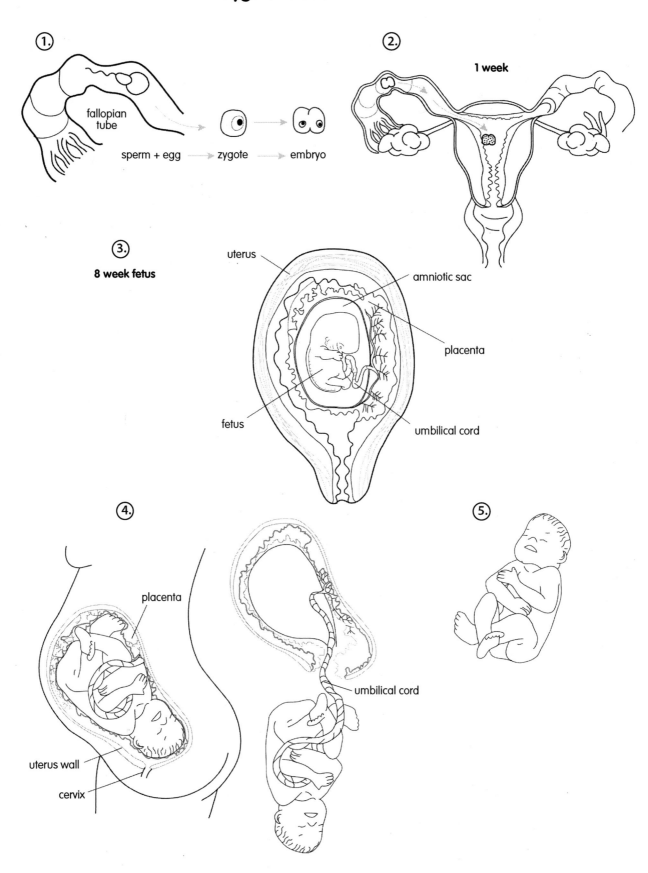

1. fallopian tube

sperm + egg ⟶ zygote ⟶ embryo

2. 1 week

3. 8 week fetus

uterus

amniotic sac

placenta

fetus

umbilical cord

4. placenta

uterus wall

cervix

umbilical cord

5.

Explore

Your Story
Chapter 16: Lab 2

Reproduction starts with fertilization and ends with birth.

The reason for the human reproductive system is to make more people. It starts with fertilization and ends with labor and delivery at a person's birth. Every mother has a labor and delivery story to tell, and she usually loves to tell it. Have you heard the story of your labor and delivery? If you have brothers and sisters, what about their labor and delivery stories? Every labor and delivery story is unique. Ask about other people's stories as well as your own so you can learn about some of the differences. For this lab you are going to hear about the labor and delivery story for your birth, and maybe the birth stories of your brothers and sisters or your parents as well.

Chapter 16: Lab *continued*

Materials

- You and a parent, grandparent, and other family members

- Paper and pen (optional)

Procedure

1. Start by asking your parent about your labor and delivery story. Have them tell you how your siblings' stories are different from your story. Then ask other mothers about their child's story. How are these stories different from your story? Have them begin with when their labor started through the delivery of the baby. Ask about where you were born—hospital, home, midwife center, elevator? Was any medication used? Who was there for the birth? Any complications? What happened after the birth? Get the whole story with all the details.

2. Optional: Write the story of your birth. If you are familiar with writing essays, then create a five-paragraph essay (introduction, conclusion, and three body paragraphs that tell the story of your birth). Add any baby pictures you have, including sonogram pictures and from right after or during your birth. Obtain a photocopy of your birth certificate and attach it to your story.

Important note: If you cannot know the story of your birth because you are adopted, or cannot otherwise speak with the person or persons who know your story, then you have two choices: You can write the story of someone else's birth, perhaps an incredible birth story you read about or saw online. Or you can create an imaginary story of your birth, taking what facts you do know and expanding upon them to create a complete, amazing story.

Absorb

Famous Doubling: Twins
Chapter 16: Famous Science Series 2

Some organisms have many babies at the same time and some only have one. Dogs and cats have litters with multiple offspring. People, horses, and elephants usually only have one baby at a time. Sometime women do give birth to multiple babies, though. When a woman gives birth to two babies, they are called twins. The twins can be either monozygotic or dizygotic. Study the image below and research twins to answer the questions on the next page.

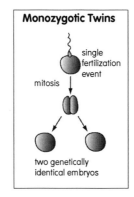

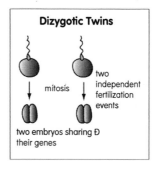

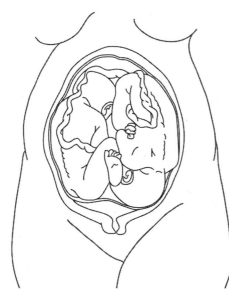

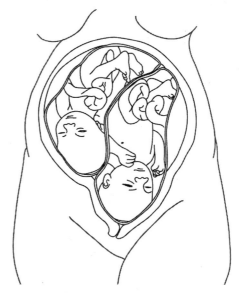

Chapter 16: Famous Science Series *continued*

How many zygotes make monozygotic twins?

Are monozygotic twins identical or fraternal?

What is the process that creates monozygotic twins?

How many zygotes make dizygotic twins?

Are dizygotic twins identical or fraternal?

What is the process that creates dizygotic twins?

Do dizygotic twins have the same DNA?

Pandia PRESS

Endocrine and Reproductive Systems
Chapter 16: Show What You Know

Multiple Choice

1. The purpose of the endocrine system is to

   ○ maintain homeostasis.
   ○ make hormones.
   ○ control when you sleep and eat.
   ○ All of the above

2. The birth process has three stages. The order they follow is:

   ○ labor, delivery of baby, delivery of placenta
   ○ delivery of placenta, labor, delivery of baby
   ○ labor, delivery of placenta, delivery of baby
   ○ delivery of baby, labor, delivery of placenta

3. Menstruation occurs when

   ○ a baby is born.
   ○ the female gamete is not fertilized.
   ○ a woman is pregnant.
   ○ the female gamete is fertilized.

4. Your endocrine system is made of

   ○ organs called receptors.
   ○ lymph nodes.
   ○ hormones.
   ○ endocrine glands.

5. An example of a homeostatic feedback mechanism is:

   ○ the heater turns on when it is hot and off when it is cold
   ○ insulin increasing glucose in your blood after you have eaten a big meal and glucagon decreasing glucose in your blood when it gets low
   ○ your olfactory nerves telling you when a food is rotten
   ○ insulin decreasing glucose in your blood after you have eaten a big meal and glucagon increasing glucose in your blood when glucose gets low

6. What is the purpose of the umbilical cord?

   ○ It transports oxygen and nutrients from the mother to the fetus, and wastes from the fetus to the mother.
   ○ It connects the fetus to the cervix.
   ○ It brings in amniotic fluid to keep waste away from the fetus.
   ○ It keeps the fetus from falling out of the mother.

Chapter 16: Show What You Know *continued*

7. Fertilization occurs in the

   ○ vagina.

   ○ testes.

   ○ fallopian tubes.

   ○ uterus.

8. The chemical messengers that send information back and forth between organs are

   ○ electrical signals.

   ○ water molecules.

   ○ vitamins.

   ○ hormones.

9. The testes are outside a male's body

   ○ so the sperm does not get too cold.

   ○ because they won't fit inside.

   ○ to keep urine away from the body.

   ○ so the sperm does not get too hot.

10. At eight weeks the embryo is called a

    ○ gamete.

    ○ zygote.

    ○ fetus.

    ○ cervix.

11. What is the purpose of the male reproductive system?

    ○ To provide genetic variability

    ○ To aid the kidneys in waste removal

    ○ To provide a place for the zygote to form

    ○ To make and deliver the male gamete

12. Females make eggs:

    ○ All the time

    ○ They are born with them all already made

    ○ Once a week

    ○ Once a month

13. An example of homeostasis is

    ○ sweating when it is hot.

    ○ hormone production.

    ○ craving sweet foods.

    ○ snoring.

14. The fetus develops in the

    ○ uterus.

    ○ vagina.

    ○ fallopian tubes.

    ○ ovaries.

15. What is the purpose of the female reproductive system?

    ○ To house and protect the female gamete

    ○ To protect and nourish the fetus

    ○ To provide a place for the zygote to form

    ○ All of the above

16. Puberty begins when the endocrine system begins making more of the hormone _____ in boys and the hormone _____ in girls.

    ○ testosterone, estrogen

    ○ human growth hormone, insulin

    ○ estrogen, testosterone

    ○ glucagon, human growth hormone

Chapter 17: Circulatory and Respiratory Systems

Read

Going Round in Circles

Chapter 17: Lesson 1 Activity

The following coloring activity is found in Chapter 17 Lesson 1 in your Textbook.

Color the Circulatory System and Parts of Blood. *As you read about the circulatory system, color the illustrations below.*

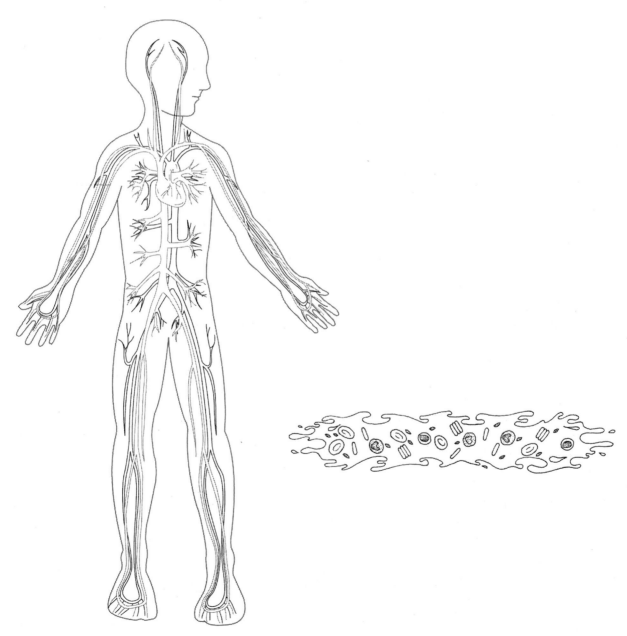

Explore

OUCH!

Chapter 17: Microscope and Lab Sheet

Name_____ **Date**_____

Today you are going to stab yourself in the name of science! Blood is very interesting to look at with a microscope. You can see red blood cells, plasma, white blood cells, and platelets.

Materials

- Needle or pin to STAB! yourself
- Slide
- Slide cover
- Microscope

- Rubbing alcohol
- Soap and water to wash your hands

Procedure

1. Wash your hands well with soap and water.

2. Pour rubbing alcohol over the needle to clean it.

3. Prick your finger with the needle. Drip a drop of your blood on a slide and cover it. Do not let the blood sample dry out. The view will differ depending on how much blood you are looking at. A light smear gives the best view of all the parts of the blood that are present.

4. Look at the drop with the microscope.
   Using the 400x magnification, draw what you see.

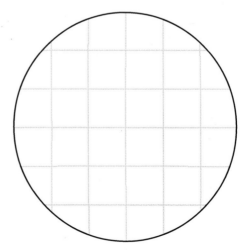

Pandia PRESS

The Beating Heart
Chapter 17: Lab 1

A stethoscope amplifies sounds from the inside of your body and makes them loud enough for the doctor to hear.

When you go to the doctor's office and the doctor puts a stethoscope in their ears and listens, have you ever wondered why? Your heart has a regular beat. It beats all day and it beats all night. You can feel it beat if you press your fingers on your chest. You can hear it beat too. Doctors use a simple instrument called a stethoscope to listen to their patients' heartbeats. A stethoscope amplifies sounds from the inside of your body and makes them loud enough for the doctor to hear. Doctors can use a stethoscope to listen and make sure your heart, stomach, and lungs are all functioning normally.

A healthy heart makes two clicking sounds for each beat as the two valves inside it open and close at different times. Doctors listen to make sure your heart sounds like that. Sometimes the heart makes an extra whooshing sound, which can indicate a heart murmur. They also listen to make sure the heart is beating at the right speed. Some health problems can cause a rapid heartbeat, which means the heart has to work harder than it should.

When a doctor asks you to breathe deeply, the doctor is listening to your lungs. The doctor is listening to hear wheezing, congestion, or breathing problems. Your stomach is working all the time. Sometimes, you can even hear it without a stethoscope. Doctors can listen to your stomach for blockages.

You are going to make a stethoscope and listen to the heart, lungs, and stomach of another person. If you have any pets who would not be stressed by your doing this, you can listen to their heart, lungs, and stomach too.

Materials

- 2 water bottles (500 ml)
- Scissors
- Duct tape

- 0.6 m vinyl tubing, with a diameter that can be fitted over the mouth of the bottle tightly, but without too much trouble

- Timer
- X-Acto knife

Chapter 17: Lab 1 *continued*

Procedure

1. Use the X-Acto knife to make a slit on both bottles about ⅔ of the way up from the bottom of the bottle. At the slit, use the scissors to cut around both bottles, cutting the bottoms from the tops. Put one end of the vinyl tubing over the mouth of each bottle. Use the duct tape where the tubing and bottle meet, to secure the tubing to the bottles. You might need to put duct tape around the cut end of the bottle, to make it more comfortable to press to your body, but this is optional.

2. Listen to your own heartbeat first. Place one end of your stethoscope over your heart and the other end up to your ear. You might need to move it around your chest to find the best place on your own body to listen to your heart. For the beats per minute, count the beats for 10 seconds and multiply by 6. Fill in the lab sheet. It might be difficult to listen to your own stomach and lungs, but you should try it.

4. When listening to another person's heart, listen first without a stethoscope, so you can hear the difference the stethoscope makes.

5. Listen to the heartbeat, breathing, and stomach of your teacher, friends, family members, and pets. Record your observations on your lab sheet. Breathing is easiest to hear from a person's back.

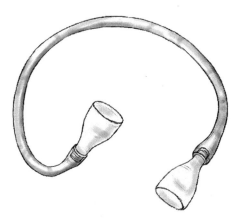

Pandia PRESS

The Beating Heart

Chapter 17: Lab 1 Sheet

Name_____ Date_____

Myself

Heart - _____ beats per minute   How it sounds:

Lungs - How they sound:

Stomach - How it sounds:

Parent/Teacher

Heart - _____ beats per minute   How it sounds:

Lungs - How they sound:

Stomach - How it sounds:

My Pet (Do you hear any differences for people and your pet?)

Type of animal:

Heart - _____ beats per minute   How it sounds:

Lungs - How they sound:

Stomach - How it sounds:

What difference did you hear with and without the stethoscope?

Make a list below for the rest of the people and pets you listened to with your stethoscope. Be sure to include your observations.

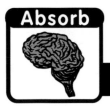

Rene-Theophile-Hyacinthe Laënnec
Chapter 17: Famous Science Series 1

Famous Inventor

What did Rene-Theophile-Hyacinthe Laënnec invent? When did he invent it?

Rene-Theophile-Hyacinthe

Laënnec's mother died when he was five years old from what disease? What does this disease do to people who have it? Was this disease serious and has a cure been found?

How did Laënnec get the idea for his invention?

How did he name his invention?

Pandia PRESS

"The Tell-tale Heart"
Chapter 17: Short Story

"The Telltale Heart" by Edgar Allan Poe

Read "The Tell-tale Heart," a short story by American writer Edgar Allan Poe. You can read the story online or from a book. After reading, explain scientifically why it was or was not possible for the old man's heart to still be beating.

Breathe In, Breathe Out
Chapter 17: Lesson 2 Activity

The following coloring activity is found in Chapter 17 Lesson 2 in your Textbook.

Color the Respiratory System. *As you read about the respiratory system, color the illustrations below.*

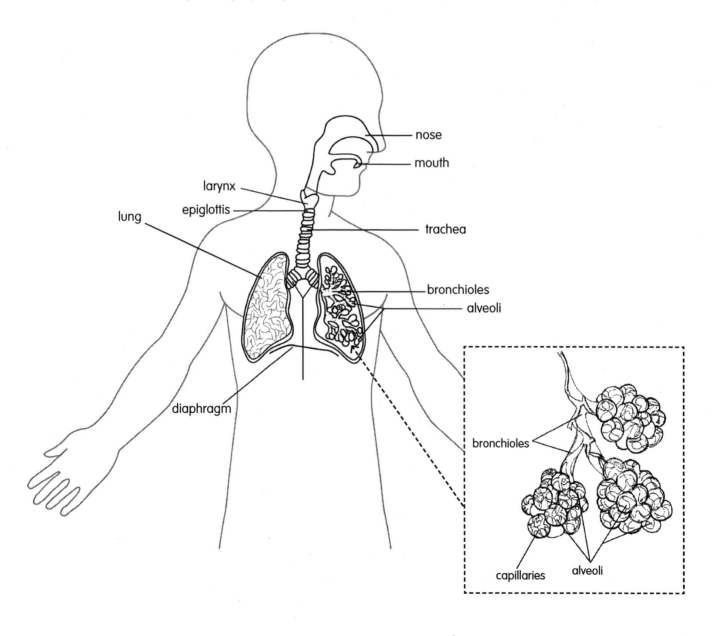

Pandia PRESS

Lung Capacity

Chapter 17: Lab 2

People come in different sizes and so do their lungs. That makes sense, because the bigger you are, the more cells you have. The more cells you have, the more oxygen you need for your cells. The amount of oxygen you take in is called your **lung capacity**. Things that affect lung capacity are regular exercise (athletes have a larger lung capacity), smoking, and diseases such as asthma, emphysema, bronchitis, and tuberculosis.

With this experiment, you will determine how much air you inhale and exhale: your **Vital Lung Capacity (VLC)** and your **breathing rate**. Your VLC measures the amount of air you exhale after you have inhaled as big a breath as you can. Your breathing rate measures the amount of breaths you take in one minute. Other lung volume measurements are the **expiratory reserve**. This is the amount of air left in your lungs after you exhale.

Materials

- Measuring stick that measures centimeters and some string, OR
- Measuring tape that measures centimeters
- Round balloon
- Calculator
- Another person

Procedure

1. Fill in the lab sheet as you go along in this experiment. Start with the first question.

2. Exhale normally, and now have your rib cage and diaphragm tighten some more; the volume of air that comes out when you do that is the **expiratory reserve**. Under normal conditions that volume of air remains in your lungs.

3. Sit and breathe normally. Have someone count the number of breaths you take in one minute. Next, run around as hard as you can for two minutes. Right after, have someone count the number of breaths you take in one minute.

Pandia PRESS

Chapter 17: Lab 2 *continued*

4. Wait until your breathing rate is normal before performing the rest of this experiment.

5. Stretch out the balloon by blowing it up five times. Rest 2 minutes after you do this.

6. Stand up straight and blow up the balloon as much as you can with ONE breath. Pinch the end of the balloon closed. Do not tie it off or let any air out. If any air escapes, redo this part of the experiment. Have the other person measure the circumference of the balloon by wrapping the string or measuring tape around the balloon at its widest point. Now you can let go of the balloon. Use the measuring stick to measure the string length in cm. Record the length. Perform this part of the experiment five times. Rest at least 2 minutes between each time.

7. Calculate your vital lung capacity. The measurement from the string or tape is the **circumference**.

Math Review

 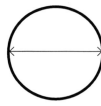 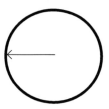

| Circumference | Diameter | Radius |

Circumference is the distance around a circle.

Diameter is the distance across a circle (from edge to edge).

Radius is the distance from the middle of a circle to its edge.

Pi or π = 3.14. Pi is the number used to calculate the circumference of a circle. It is the ratio of the circumference of a circle to its diameter. Pi is always the same number for every circle.

Formulas:

• Circumference or **C = 2 π r**

> *The circumference of a circle is 2 x 3.14 x the radius.*

• Radius or **r = C/(2 π)**

> *The radius of circle is the circumference divided by two times Pi.*

• Volume or **V = ⁴⁄₃ π r³**

> *The volume of a sphere (the amount of space inside, or its capacity) is ⁴⁄₃ x 3.14 x the cube of its radius. This is a cubic measurement (three-dimensional). So write it in milliliters (ml). (A milliliter is a centimeter cubed.)*

• Average Volume = <u>**V #1 + V #2 + V #3 +V #4 + V #5**</u>
  5

> *Volume from the five trials added together and then divided by 5.*

Lung Capacity

Chapter 17: Lab 2 Sheet

Name_____ Date_____

Breathing Rate

Will there be a difference in your resting and running breathing rate? Why or why not? If yes, how do you think it will change? Why?

Resting breathing rate: _____breaths/minute

Running breathing rate: _____breaths/minute

Change in your breathing rate:

_____ - _____ = _____

(Running breathing rate)   (Resting breathing rate)

Lung Capacity

Data Table

	circumference in cm, C	radius in cm, r = C/(2π)	volume in mL, V = ⁴⁄₃ π r³
1			
2			
3			
4			
5			

My average volume = average Vital Lung Capacity, VLC = _____mL

Chapter 17: Lab 2 Sheet *continued*

Questions

Were you surprised by the expiratory reserve in your lungs? Why or why not?

Why did you take five measurements?

How do you think VLC would change for someone who smoked? Why?

How do you think VLC would change for someone who was a marathon runner? Why?

Pandia PRESS

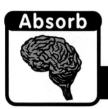 **Absorb**

Sherpas

Chapter 17: Famous Science Series 2

Famous Lung Capacity

Where do Sherpas live?

Where does the name *Sherpa* come from?

What mountain do Sherpas help people climb, and how do they help them climb it? How high is this mountain?

One of the problems facing climbers is the concentration of oxygen. While the percentages of molecules in the air are the same on the top of Mount Everest as they are at sea level (21 percent oxygen, 78 percent nitrogen, and 1 percent other gases), the concentration of these molecules is lower. Concentration is a measure of the amount of particles in a given volume. For example, if you put 1 teaspoon of sugar into 1 liter of iced tea and your sister puts 5 teaspoons of sugar into 1 liter of iced tea, the concentration of sugar in your sister's iced tea would be 5 times greater than the concentration of sugar in your iced tea. How much less is the concentration of oxygen molecules at the top of Mount Everest?

Why is this a problem?

Sherpas have a larger lung capacity than people who live at lower elevations. How does this help them deal with lower levels of oxygen?

Circulatory and Respiratory Systems

Chapter 17: Show What You Know

Multiple Choice

1. Your arteries

    ○ carry blood away from your heart to the rest of your body.
    ○ carry blood to your heart from your lungs.
    ○ are one cell thick, which allows transport through the them.

2. Your capillaries

    ○ carry blood away from your heart to the rest of your body.
    ○ carry blood to your heart from your lungs.
    ○ are one cell thick, which allows transport through them.

3. Your veins

    ○ carry blood away from your heart to the rest of your body.
    ○ carry blood to your heart from your lungs.
    ○ are one cell thick, which allows transport through the them.

4. Mucus

    ○ is the waste product of bacteria.
    ○ is filled with plasma and platelets.
    ○ traps germs so they don't get into your lungs.
    ○ lubricates your nostrils so air can get into your lungs.

5. Your alveoli and capillaries work together to

    ○ produce mucus.
    ○ transport oxygen into your lungs and carbon dioxide into your blood.
    ○ transport carbon dioxide into your lungs and oxygen into your blood.
    ○ help you breathe.

6. The muscle under your lungs that helps you breathe is your

    ○ larynx.
    ○ trachea.
    ○ bronchi.
    ○ diaphragm.

Chapter 17: Show What You Know *continued*

7. The protein molecule that carries oxygen in your red blood cells is

   ○ hemoglobin.
   ○ plasma.
   ○ platelets.
   ○ carbon dioxide.

8. _____ rush to the site of a cut and produce _____ ,which forms a protective covering over the cut.

   ○ Platelets, fibrin
   ○ Plasma, white blood cells
   ○ Red blood cells, hemoglobin
   ○ White blood cells, fibrin

9. The tubes that go into your lungs are called

   ○ alveoli.
   ○ trachea.
   ○ bronchi.
   ○ larynx.

10. Plasma in your blood carries

    ○ food.
    ○ hormones.
    ○ waste products.
    ○ All of the above

11. Your heart has two sides that pump blood:

    ○ the left side pumps blood out and the right side pumps blood in
    ○ the right side pumps blood out and the left side pumps blood in
    ○ both pump out and then in at the same time
    ○ both pump in and then out at the same time

12. The part of your brain that controls your heartbeat and your breathing is your

    ○ cerebrum.
    ○ pancreas.
    ○ cerebellum.
    ○ medulla.

13. Vocal cords are two folds of tissue that stretch across this organ, allowing you to talk. Your vocal cords are in your

    ○ trachea.
    ○ bronchi.
    ○ larynx.
    ○ alveoli.

14. When you eat and drink, this closes over the entrance of your trachea so no food or drink gets into your lungs:

    ○ bronchi
    ○ epiglottis
    ○ larynx
    ○ alveoli

15. Your skin and white blood cells fight infections together by:

    ○ white blood cells form a protective layer on your skin
    ○ your skin funnels germs through your pores to your white blood cells, so they can destroy them
    ○ your skin prevents most germs from getting in, but when they do, the white blood cells destroy them
    ○ when you get a cut, white blood cells rush to the site to form a clot over the cut

Pandia PRESS

Chapter 18: Skeletal and Muscular Systems

A Structured Life

Chapter 18: Lesson 1 Activity

The following coloring activity is found in Chapter 18 Lesson 1 in your Textbook.

Your Skeletal System. *As you read about the anatomy of the knee, color the illustration below.*

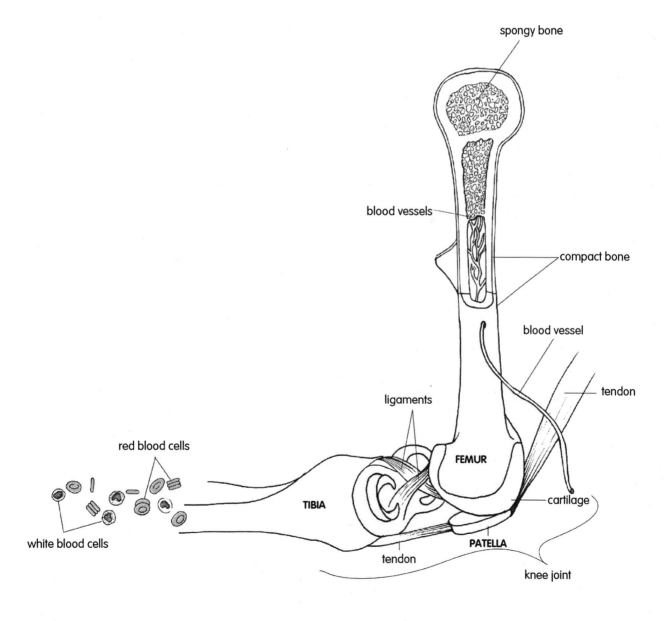

Your Skeletal System. *As you read about the skeleton, color the illustration below and color code the key.*

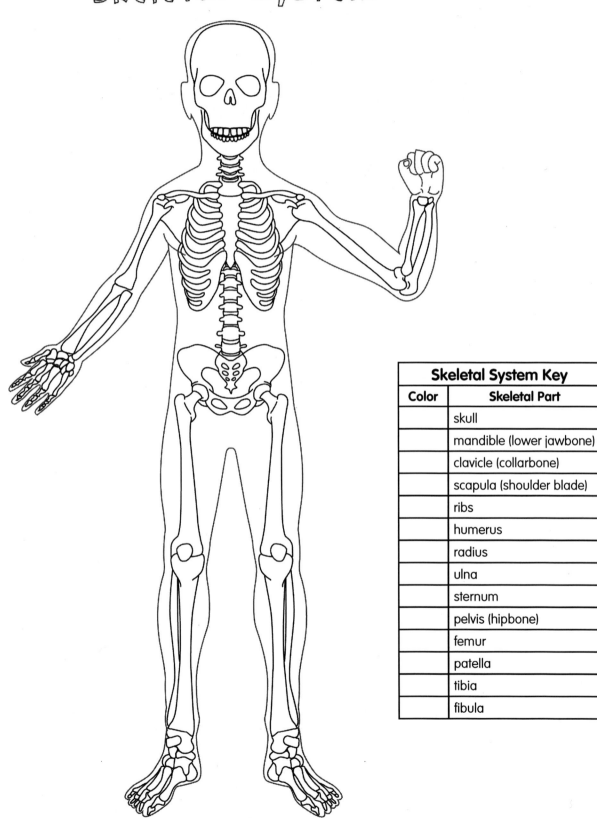

Skeletal System Key	
Color	**Skeletal Part**
	skull
	mandible (lower jawbone)
	clavicle (collarbone)
	scapula (shoulder blade)
	ribs
	humerus
	radius
	ulna
	sternum
	pelvis (hipbone)
	femur
	patella
	tibia
	fibula

The Joint Detective Game
Chapter 18: Lab

Your bones meet at joints. Joints are made from ligaments, which hold the bones together, and cartilage. Cartilage is very slippery. It reduces friction at the joint, where the bones rub against each other.

There are three types of joints in your body: immovable, slightly movable, and freely movable. Immovable joints are in your skull at the places the bones protecting your brain meet. The joints in your ribs are slightly movable. The rest of your joints are freely movable; that does not mean they can move any direction at all, though. It does mean that with the help of your muscles you can get up and dance, play basketball, or move about freely.

Joints are all around you, not just in your body. There are six different types of movable joints in your body. Do you think you can identify the types of joints in your body? Do you think you can find things in your house, school, or car that move like each joint type? Why don't you try it and find out?

Six Types of Joints

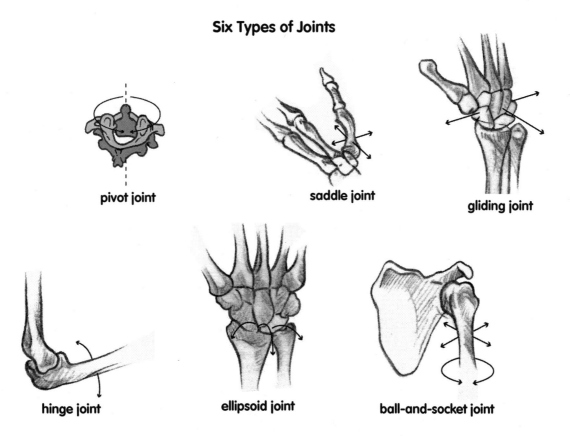

pivot joint

saddle joint

gliding joint

hinge joint

ellipsoid joint

ball-and-socket joint

Chapter 18: Lab *continued*

Materials

- Your house, school, and/or car
- Your body
- Game sheet

Procedure

Number of Players: 1 to as many as can be found. Each player should have a copy of the Game Sheet.

The Rules

- Starting with your toes, work your way up the different joints on your body. Using the illustrations on the previous page, try to identify each type of joint in your body. Then look around your house, school, and/or car and find things that mimic the movement of the different joints in your body. Write the name of the first joint you identify and any others of the same type in the correct box on the chart on the Game Sheet. For example, you might identify your toe as a hinge joint, and in your house you find a book that opens and closes like a hinge joint. Record duplicate objects and body parts only one time. For example, you most likely have ten toes that are hinge joints, but just write "toes" one time. And there are hundreds of books in your house (perhaps thousands), but sorry, they would count as only one identification.

- Don't worry about the joints on your body that do not move (like those in your skull).

The Points

- Give yourself 10 points when you first identify a type of joint. That's 10 points for your body and 10 for your house/school/car. That means there are 120 points available for the first identification of the six types of joints.

- Bonus points: 2 points for all other examples of that joint type. 1 point if you had to look up the joint type in a book or on the Internet, but at least you know for sure.

- Take away 2 points for misidentifying a joint type. If you misidentify the first joint you choose for a type, you get -2 for that too. (If there is another joint of that type that you correctly identify, you can give yourself +10 for that joint identification.)

Scoring

100 + points: Have you considered a career as an osteopath?

70 to 99 points: You are a super sleuth joint detective!

40 to 69 points: Fairly good work.

39 or less: Do you have joints? Are you from Earth?

The Joint Detective Game

Chapter 18: Game Sheet

Name_____ Date_____

The Joint Detective Game

BODY	Ball-and-socket joint	Hinge joint	Pivot joint	Saddle joint	Gliding joint	Ellipsoid joint
First identified joint of this type. 10 points each						
Other identified joints of this type. 2 points each						
HOME	Ball-and-socket joint	Hinge joint	Pivot joint	Saddle joint	Gliding joint	Ellipsoid joint
First identified object of this type. 10 points each						
Other identified object of this type. 2 points each						

Add up the points up here:

I Can't Sit Still

Chapter 18: Lesson 2 Activity

The following coloring activity is found in Chapter 18 Lesson 2 in your Textbook.

Your Skeletal System. *As you read about your muscular system, label and color the illustrations.*

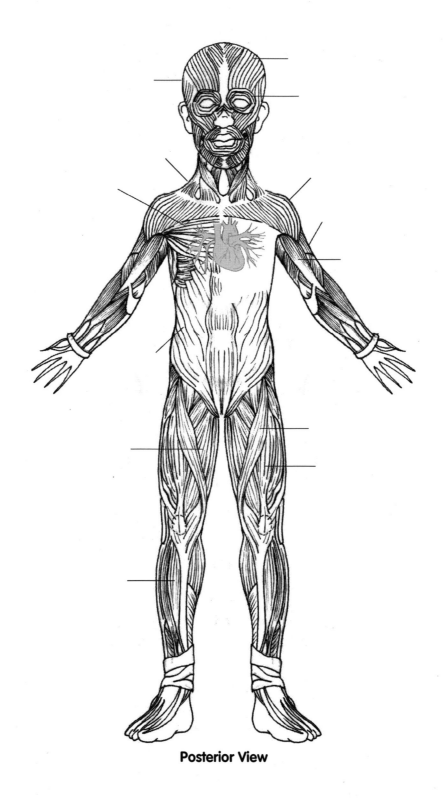

Posterior View

Pandia PRESS

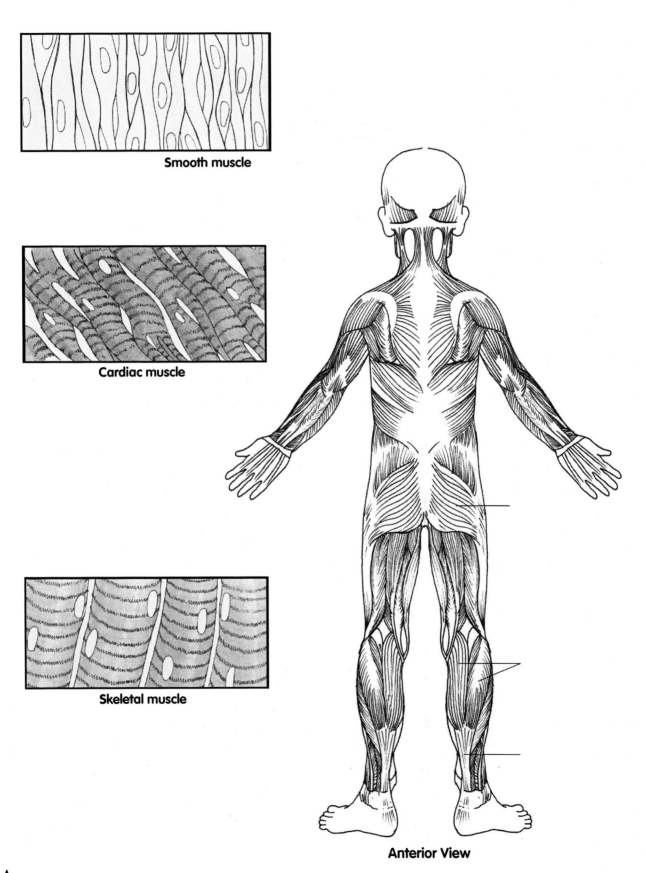

Smooth muscle

Cardiac muscle

Skeletal muscle

Anterior View

A Chicken Wing Thing
Chapter 18: Dissection and Microscope Lab

Your muscles and bones work together to make you move. The muscles and bones of other organisms do the same thing. Today you will dissect a chicken wing and look at its bones, muscles, and joints to see how they work together.

Materials

- Book, any of medium weight
- Chicken wing
- Surface to dissect the chicken wing

- Paper towels
- Scissors
- Scalpel or paring knife
- Gloves

- Magnifying glass (optional)
- Muscular and skeletal system illustrations from Chapter 18 in your textbook.

Optional Microscope Materials

- Microscope
- Slide
- Slide cover

- Water
- Syringe
- Cutting board

- Freezer
- Desk lamp or other light source
- Methylene blue

Procedure

Dissection

- Raw chicken can carry Salmonella bacteria. Do not touch your mouth with your hands or the tools during this lab.

- For the microscope portion of this experiment, you need to save samples of material as you are dissecting the chicken wing. Save the following:

| Fat | Muscle | Blood vessel | Tendon | Ligament |
| Cartilage | Bone | Skin | Bone marrow | |

As you dissect these, put them in the freezer. It is easier to get a thin slice if they are partially frozen.

Chapter 18: Dissection and Microscope Lab *continued*

1. Examine your own arm. Rest it on a table, fully extended. Feel the muscles above and below your elbow. Tighten your muscles, now feel them. Relax your muscles and feel your upper and lower arm again. With your elbow resting on the table, bend your arm up and down. Flatten your hand and put a book on it. Lift the book with your arm. Feel the muscles in your upper and lower arm with the book raised. You are feeling your muscles work together by pulling on the bones in your arm to make it move. Answer question 1 on the lab sheet.

2. Examine the chicken wing. Look carefully at the skin. Look at the shoulder where the wing was cut from the chicken's body. Do you see cartilage, bones, or marrow? Record your observations. Answer question 2 on the lab sheet.

3. Tug at the chicken wing. Observe how the joints move. Locate the upper arm, lower arm, and wingtip.

4. With the scissors and scalpel, carefully cut away the skin from the chicken bone. Cut down the middle starting at the shoulder and down toward the wingtip. Do not cut the muscles or joint. Chicken skin can be difficult to cut away; take your time and slowly peel the skin with the tissue away from the chicken. Answer question 3 on the lab sheet.

5. Fat, a yellow slimy material, is under the skin: find and identify some.

6. The muscles have the same names in a chicken wing as they do in your arm. Refer to the muscular system illustrations in Chapter 18 to label the *muscles* of the upper part of the chicken wing on Diagram 1 on the lab sheet.

7. Hold the wing down by its shoulder. Pull on the triceps muscle. Pull on the biceps muscle. Answer question 4 on the lab sheet.

8. Find the shiny tendons at each end of the muscles. Pull on the tendons at the elbow joint. Pull on the tip of the wing. What happens when you do this? Label the *tendons* on Diagram 1. Answer question 5 on your lab sheet.

9. Find the thin, brown strand and pull it aside. This is a blood vessel. Answer question 6 on the lab sheet.

10. With the scalpel, carefully remove the muscles from the bones one at a time. You do this by cutting the tendon and peeling the muscle from the bone. As you remove each muscle, examine how this changes the ability of the wing to move. After you have cut away one muscle and looked at how this affects movement, answer question 7 on the lab sheet.

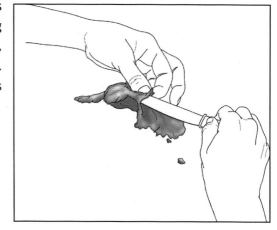

11. Cut away all the muscles.

Chapter 18: Dissection and Microscope Lab *continued*

12. The bones have the same names in a chicken wing as they do in your arm. Refer to the skeletal system illustrations in Chapter 18 to label the **bones** of the chicken wing on Diagram 2 on the lab sheet.

13. Look at the elbow joint after the muscle has been removed. Ligaments connect bones where they meet at joints. Label the **ligaments** and **joints** on Diagram 2 on the lab sheet.

14. Move the bones around at the elbow joint. Observe the range of motion of the elbow joint. Answer question 8 on the lab sheet.

15. With the scalpel, carefully cut the ligaments at the elbow joint. Separate the elbow. Look at the slippery material between the joint. This is cartilage. Label the **cartilage** on Diagram 2. Answer question 9 on the lab sheet. Carefully separate the ulna and the radius. Observe how they all fit together puzzle-like.

16. Carefully, cut the bone in half with the scissors or break it in half so that you can look inside it. Answer question 10 on the lab sheet.

Microscope

1. Looking at the parts:

   - One at a time, on a cutting board, put a piece of slightly frozen muscle, tendon, cartilage, ligament, or skin. The fat is done as a smear and does not need to be cut. Get as small a piece of bone as possible. Scissors, pliers, or even a hammer can help get this sample.

   - With the scalpel, CAREFULLY (do not cut yourself) cut a thin slice.

   - Make a wet mount slide with the slice. Stain all the samples with methylene blue, EXCEPT the bone marrow and the blood vessel.

   - Look at it under the microscope. It depends on the sample as to which magnification gives the best view. Be careful not to dirty your lens by touching it to any of the samples.

   - Draw a view of each.

2. For the bone marrow, scrape out some bone marrow from the bone that was broken, smear it on a slide, add a little water and a slide cover, and look at bone marrow. Draw the view.

3. For the bone sample, make a dry mount slide without a slide cover. View this slide with 40x magnification. Draw the view.

A Chicken Wing Thing

Chapter 18: Dissection Lab Sheet

Name_____ Date_____

Questions

1. Describe what happened in your muscle and skeletal systems when you raised the book.

2. What do you see at the cut-away shoulder joint? Do you think the chicken wing came from the right or left side of the body? Why?

3. Describe the chicken skin. The skin is a part of what organ system?

   • When you cut away the skin, there was tissue connecting the skin to the muscle underneath it. What is this type of tissue called?

   • What is the name of the layer of skin on the outside of the chicken wing?

   • What is the name of the skin on the inside that is connected to the muscle?

4. Skeletal muscles always come in pairs, called **antagonistic pairs**. In order to move the bone, each muscle moves in the opposite direction. One of the muscles is a flexor muscle. It flexes or bends the joint. The other muscle is an extensor muscle. It extends or straightens the joint. Which muscle in the upper arm of the chicken wing is a flexor and which is an extensor? Add the names of the muscles as labels to Diagram 1.

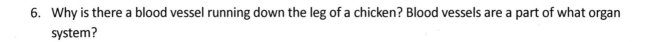

Chapter 18: Dissection Lab Sheet *continued*

5. Why are tendons attached to the ends of the muscles? Tendons are made from one type of tissue. What type do you think that is? Why?

6. Why is there a blood vessel running down the leg of a chicken? Blood vessels are a part of what organ system?

7. Describe how movement was affected when one muscle in a pair was cut.

8. What kind of joint is the elbow joint? How does this compare with your elbow joint?

9. How do the three bones at the elbow joint fit together? You can answer this with words or a labeled drawing.

Pandia PRESS

Chapter 18: Dissection Lab Sheet *continued*

10. Describe with words, a drawing with labels, or a labeled photograph, the inside of the bone. What is the name of the soft, red material inside the bone? What important function does this material provide?

11. **Microscope:** Compare the shape of cheek cells, red blood cells, striated muscle cells like those that make up the arm muscles, and the cells that make up tendons. How does form fit function? Now that you have looked at a blood vessel, why do you think red blood cells have the shape they do?

Chapter 18: Dissection Lab Sheet *continued*

Diagram 1: The Chicken Wing with Muscles

Label the muscles and tendons, biceps, triceps, and tendons at elbow.

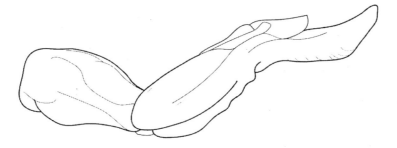

Diagram 2: The Chicken Wing with No Muscles

Label the bones, ligaments, and joints: shoulder joint, humerus, elbow joint, wrist joint, radius, ulna, ligaments at elbow, and cartilage.

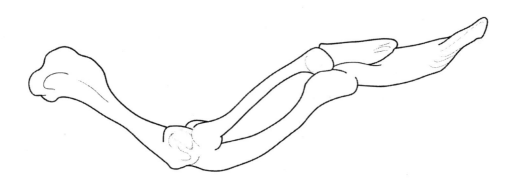

A Chicken Wing Thing

Chapter 18: Microscope View Sheet

Name_____ **Date**_____

Specimen _____ Type of mount_____

Type of stain used_____

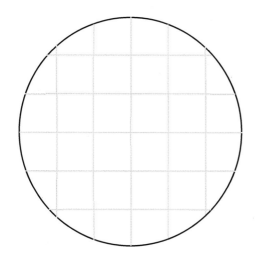

Fat, magnification _____ x

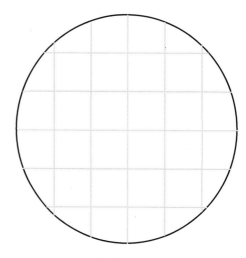

Muscle, magnification _____ x

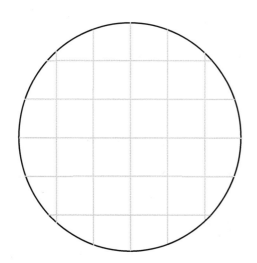

Cartilage, magnification _____ x

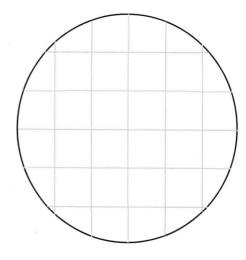

Tendon, magnification _____ x

Comments:

Chapter 18: Microscope View Sheet *continued*

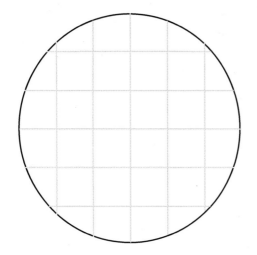

Blood Vessel, magnification _____ x

Ligament, magnification _____ x

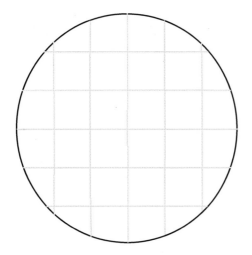

Bone Marrow, magnification _____ x

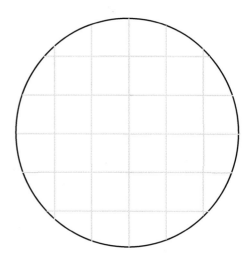

Bone, magnification _____ x

Comments:

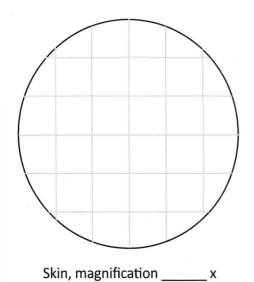

Skin, magnification _____ x

Pandia PRESS

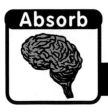

Absorb

Perfecting Prosthetics

Chapter 18: Famous Science Series

Famous Inventions

What does the word *prosthetic* mean?

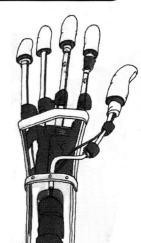

How long have people been using them? What were they made from in ancient times?

What four organ systems have to be coordinated to make a limb move?

It would be really hard to design something that integrates all these organ systems into one package. Reach out in front of you and pick up something as delicately as you can. Now think about everything that went into your picking up that object. Can you imagine what it would entail to build an artificial limb that could do all that? An agency in the United States military called Defense Advanced Research Projects Agency (DARPA) did just that. The program is called Revolutionizing Prosthetics.

Skeletal and Muscular Systems

Chapter 18: Show What You Know

Multiple Choice

1. How does shivering help your body maintain homeostasis?

   ○ When you are cold, your muscles generate heat by shivering.
   ○ Shivering helps calm you down.
   ○ When you shiver, you make more white blood cells.
   ○ Shivering stimulates your endocrine system to make more hormones.

2. Skeletal muscles are

   ○ involuntary muscles.
   ○ cardiac muscles.
   ○ voluntary muscles.
   ○ smooth muscles.

3. Involuntary muscles move

   ○ when your brain tells them to move.
   ○ without you thinking about them moving.
   ○ to make you walk.
   ○ All of the above

4. Voluntary muscles move

   ○ when your brain tells them to move.
   ○ without you thinking about them moving.
   ○ to make your heart beat.
   ○ All of the above

5. When red blood cells are made, they head to your

   ○ cells.
   ○ heart.
   ○ lungs.
   ○ brain.

6. What organ is made of cardiac muscles?

   ○ Liver
   ○ Heart
   ○ Lungs
   ○ Brain

7. Your muscles are attached to bones by

   ○ cartilage.
   ○ tendons.
   ○ skin.
   ○ ligaments.

8. Your bones are attached to each other by

   ○ cartilage.
   ○ tendons.
   ○ skin.
   ○ ligaments.

Pandia PRESS

Chapter 18: Show What You Know *continued*

9. The slippery material at joints where two bones meet is called

   ○ cartilage.
   ○ tendons.
   ○ skin.
   ○ ligaments.

10. Smooth muscles

    ○ line blood vessels.
    ○ are attached to skin and bone.
    ○ are also called striated muscles.
    ○ make muscles move smoothly.

11. The process where cartilage parts turn into bone as you age is called

    ○ hardening.
    ○ tendonitis.
    ○ mitosis.
    ○ ossification.

12. The outer covering on bones is called

    ○ spongy bone.
    ○ compact bone.
    ○ femur.
    ○ patella.

13. The webbed part of bone that allows compression and absorbs force is called

    ○ spongy bone.
    ○ compact bone.
    ○ femur.
    ○ patella.

14. The point where two bones meet is a

    ○ femur.
    ○ patella.
    ○ joint.
    ○ ligament.

15. Blood cells are made in

    ○ compact bone.
    ○ spongy bone.
    ○ the heart.
    ○ muscles.

Chapter 19: Immune and Lymphatic Systems

Read

The Warrior Systems

Chapter 19: Lesson 1 Activity

The following coloring activity is found in Chapter 19 Lesson 1 in your Textbook.

Immune and Lymphatic Systems. *As you read about the immune and lymphatic systems, color and label the illustration below.*

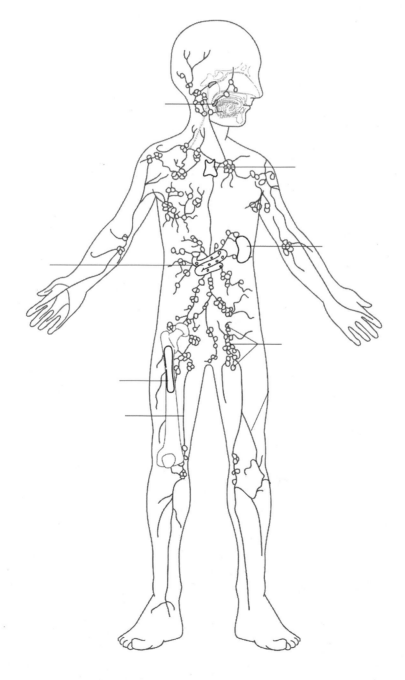

Explore

Bacteria Out of Control

Chapter 19: Lab 1

Bacteria infect an organism. The immune system does battle, conquering the bacteria.

After that, the organism's immune system recognizes that type of bacteria and it cannot make that organism sick anymore.

Your immune and lymphatic systems work hard to keep you healthy. They constantly scour your body looking for pathogens. Despite all their hard work, sometimes you still get sick. The war between the immune and lymphatic systems and bacteria has been going on for millions of years. Bacteria have an important weapon in their arsenal: the ability to reproduce quickly. This is one of the advantages for organisms, like bacteria, that reproduce asexually.

If conditions are ideal, like the warm and moist inside of your body, the average time for bacteria to reproduce (divide) is 15 minutes. One bacterium divides into two, 15 minutes later those two bacteria divide in two, making four bacteria; and then those four bacteria divide in two, making eight bacteria. It takes less time than you would think to have over 1,000 bacteria. Today you will see just how long that takes.

It usually takes one or more cells from your immune system to deal with one cell of an invader. When bacteria are multiplying so rapidly, it is a lot of work for your body to keep up and kill the bacteria. Your body responds with fevers and by making more white blood cells.

Today's lab will give you an idea what your immune and lymphatic systems have to contend with, dealing with an invader that can reproduce so quickly. Today's bacteria are cleverly disguised as a piece of paper, "bacteria papericium." With it, you will discover how fast real bacteria can multiply just by doubling every 15 minutes.

Chapter 19: Lab *continued*

Materials

- 8½"x 11" piece of plain paper
- Calculator

Procedure

1. In the middle of the piece of paper write *bacteria papericium*. This will be the original parent organism, the infecting cell of bacteria.

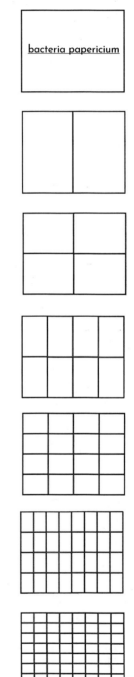

2. Take a look at the graph on the lab sheet. The X-axis (going across from left to right) is the generation axis. G1 is the first generation, G2 is the second generation, G3 is the third generation, and so on up to G11, which is the eleventh generation. Each time *bacteria papericium* reproduces, which occurs each time you rip the paper "organisms" in half, it is a new generation. The Y-axis (moving up the graph from the bottom to the top) is the number of organisms in that generation. At G11 you will have halved each piece of paper ten times (not eleven because G1 starts with the first piece of paper).

3. Start by counting the parent organism (the entire piece of paper) which equals 1. Make a dot on the graph at this point. The point is at (G1, 1) like this:

Note: The points you will graph from G1 through G4 are very close to the X-axis. 1 is slightly above the 0 line. You do not need to worry about being exact with where you place your point. An approximate placement is good enough for this lab.

4. Fold the paper in half, unfold it, and rip it in two pieces along the fold line. This is the second generation. Graph this point (G2, 2).

5. Fold the two pieces of paper in half, unfold, and rip to make four *bacteria papericium*. This is the third generation. Graph this point (G2, 4).

6. Fold, rip, and record four more times and graph up to the seventh generation.

7. At this point you can stop folding and ripping the paper and use a calculator to determine the number of individuals in G8 through G11. To do this, multiply the number of organisms in the previous generation by 2, since they double every generation. Graph each generation.

8. Connect the dots on the graph.

9. Answer the questions on the lab sheet.

Explore

Bacteria Out of Control

Chapter 19: Lab Sheet

Name_____   Date_____

Bacteria Papericium Out of Control

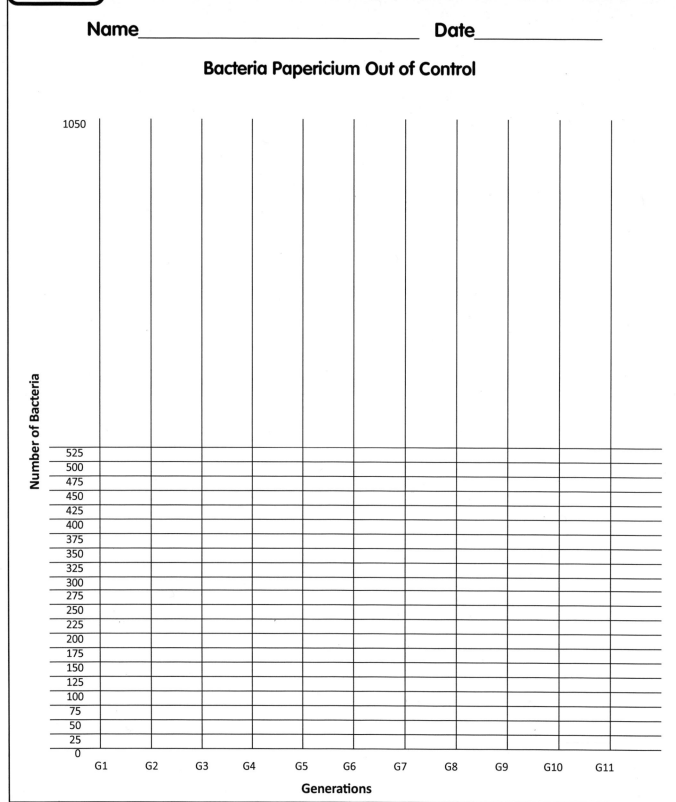

Number of Bacteria

1050

525
500
475
450
425
400
375
350
325
300
275
250
225
200
175
150
125
100
75
50
25
0

G1   G2   G3   G4   G5   G6   G7   G8   G9   G10   G11

Generations

Chapter 19: Lab Sheet *continued*

Questions

1. If it takes bacteria 15 minutes to divide, how long would it take for 11 generations to go from G1 to G11?

2. One bacterium can become 1 million bacteria in seven hours. How does that make it hard for your immune system to respond?

3. This type of growth is called ***exponential growth***. Look up *exponential growth* in the dictionary and write the definition here.

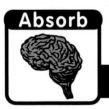

Absorb

Smallpox: What Columbus Brought
Chapter 19: Famous Science Series

Are you caught up on your immunizations? Most children in the world today are immunized against several different diseases with vaccines. Vaccines contain a neutralized form of the pathogen. They expose your immune system to the pathogen. After that, your body recognizes that pathogen and you are immune to the disease it causes. That means if you are exposed to the pathogen causing that disease, you will not get sick from it.

There have been a few types of pathogens that over the course of human history have caused much death and suffering. The smallpox virus is one of those. What happens to people infected with it?

Chapter 19: Famous Science Series *continued*

When and where did smallpox originate?

Was smallpox a problem in Europe too?

Scientists believe smallpox originated in India or Egypt over 3,000 years ago. It was a devastating disease that killed an average of 30 percent of the people who got it. Explorers and settlers brought it with them to the Americas. Indigenous people in the Americas had never been exposed to the virus before, and had little or no immunity to it.

How and when did smallpox come to the Americas?

What percentage of Native Americans are thought to have been killed by the virus?

How did this virus help bring down the Aztec Empire?

Smallpox is considered to be an eradicated virus. What does that mean?

How was smallpox eradicated?

Organ Systems Work Together
Chapter 19: Activity

Organ Systems Work Together to Make Organisms

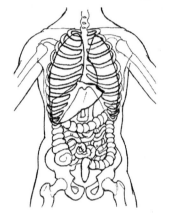

Organ systems can be studied separately but they do not function separately. So far, you have looked at how organ systems work separately or in tandem with one other system. In the last section of this unit, you will look at how all the systems work together.

Below is a list of the twelve organ systems followed by twelve descriptions. Each description relates how one of the organ systems works with the other eleven. Match the description with the name of the organ system.

Nervous System	Integumentary System	Immune System
Digestive System	Muscular System	Lymphatic System
Urinary System	Endocrine System	Reproductive System
Respiratory System	Skeletal System	Circulatory System

1. This system takes food and turns it into molecules your cells need to grow and to make energy. Since all the organs in your body are made from cells, this system powers the other eleven organ systems. This system works directly with your circulatory system when blood carries food molecules to your cells.

2. This system controls the amount of water in your body. Chemical reactions in your body occur in water. Chemical reactions occur in every organ system in your body. Therefore, this system works with every organ system in your body to make sure they have the water they need. This system also removes waste generated by the other organ systems.

3. This system connects your brain to the other organs in your body. It controls the functioning of all the other organ systems.

Pandia PRESS

Chapter 19: Activity *continued*

_____ 4. This system is in charge of maintaining homeostasis. It works closely with the circulatory system, making and sending hormones to the cells in your body. Hormones tell the cells in the organs of each organ system what to do to keep your body within the narrow ranges of homeostasis. Every system in your body participates in maintaining homeostasis; therefore, every system works with this one.

_____ 5. This system connects the organs of your immune system. It works with your immune system to keep your other organ systems from getting sick.

_____ 6. This system connects all the organ systems to one another. It exchanges oxygen and carbon dioxide with your respiratory system for the other systems. It transports food and water molecules for your digestive and urinary systems to the other systems. It carries hormones for your endocrine system to the other systems.

_____ 7. This system gives you and your organ systems support and protection. This system is necessary for motion, which allows you to get things you need like food and water. This system is where blood cells are made. Red blood cells are the part of your blood that carries oxygen and carbon dioxide back and forth to and from your other systems. Your white blood cells are the pathogen fighters in your body.

_____ 8. This system protects all the other systems by enclosing them in a protective case. It helps maintain body temperature and water homeostasis, which is important to all your organ systems.

_____ 9. Without this organ system you would not be. It is because of this organ system that you have the other eleven organ systems.

_____ 10. Every cell in every organ in every organ system relies on this organ system. No cell, tissue, organ, or organ system can function without oxygen or with too much carbon dioxide. This system takes oxygen into your body and lets carbon dioxide out of it. Your circulatory system takes care of the rest.

_____ 11. Every organ system in your body is counting on this organ system to keep them healthy by protecting them from pathogens.

_____ 12. This organ system is all about movement. Your brain says go, and this organ system does. It moves the other organ systems where they need to go to get what they need to function.

Explore
Putting Them All Together
Chapter 19: Lab 2

All your systems are working together right now to make you the organism.

You are an organism made of organ systems. Your organ systems are made up of organs. All your organs are made from two or more types of tissue. The tissues are all made of cells, all of which have your own unique DNA inside of them. Today you will focus on how your organ systems work together.

Materials

- You and another person
- Outdoors

- Paper

Chapter 19: Lab 2 *continued*

Procedure

1. Go outside. Relax and think about your organ systems. Answer the following questions orally to another person.

2. What is your nervous system doing right now? Include what your senses of sight, hearing, and smell are doing. How does your nervous system interact with your other systems?

3. What is your integumentary system doing right now? What are you touching and feeling? Your body is being bombarded with molecules in the air—are they getting inside you? How does your integumentary system interact with your other systems?

4. What did you eat and drink today? Your digestive and urinary systems are hard at work. Explain how. How does your digestive and urinary system interact with your other systems?

5. If you are alive, your circulatory system must be working. What is it doing right now? Can you feel it? How does your circulatory system interact with all your other systems?

6. How is your respiratory system working with your circulatory system right now? You are not thinking about your heart beating, your lungs breathing, or your blood circulating. Which system is controlling these functions? What is the name of the specific part of this system that controls these things? How does your respiratory system interact with your other systems? How does your respiratory system interact with the other organisms that are outdoors?

7. Is your body in a state of homeostasis right now? Your endocrine system works with the rest of your body to maintain homeostasis. How is your heart rate? Are you hungry? Do your cells have enough energy to do what you want to do? Your endocrine system controls these. How does your endocrine system interact with all your other systems?

8. Think about your skeleton. Touch your skull, then feel the bones in your pinky finger. Why do you have so many different sizes and shapes of bones? How does your skeletal system interact with your other systems?

9. Rotate the wrists of both hands. When you move, your nervous system, integumentary system, digestive system, urinary system, circulatory system, respiratory system, endocrine system, skeletal system, and muscular system all coordinate to make that movement happen. Explain how.

10. Your lymphatic system and immune system must be hard at work or you would be too sick to be outside doing this lab. Explain how they are working together with the rest of your systems so you can do science.

11. Get up, run around, and explain to the other person how and which organ systems are working together so you can run, jump, play, and talk.

12. Optional: Orally or in writing, observe another organism (animal or plant) and relate how its organ systems are working together.

Immune and Lymphatic Systems
Chapter 19: Show What You Know

Multiple Choice

1. Mucus and stomach acid prevent pathogens from getting into the rest of your body. What else is a barrier for infection?

   ○ Spleen
   ○ Skin
   ○ White blood cells
   ○ Thymus

2. The first time your body encounters a pathogen,

   ○ the pathogen has time to multiply, until your immune system figures out how to kill it.
   ○ the pathogens begin killing each other.
   ○ white blood cells begin attacking each other.
   ○ your lymph becomes infected.

3. Vaccines work by

   ○ killing pathogens.
   ○ attacking lymphocytes.
   ○ producing lymph.
   ○ exposing your immune system to a less toxic form of the pathogen.

4. To be immune means

   ○ your body recognizes a pathogen and because of that it can no longer make you sick.
   ○ the pathogen can no longer get into your body.
   ○ you will not have a fever if you get sick.
   ○ the disease is caused by bacteria, not a virus.

5. The main function of memory lymphocytes is

   ○ destroying pathogens.
   ○ remembering to make more lymph.
   ○ remembering pathogens your body has encountered before.
   ○ remembering to make white blood cells when you have an infection.

6. White blood cells kill pathogens by

   ○ taking them to red blood cells, which carry them to your kidneys.
   ○ filtering them out in your lymph nodes.
   ○ infecting them.
   ○ engulfing them or breaking them apart.

Chapter 19: Show What You Know *continued*

7. When pathogens make you sick, you can get a fever. What causes fevers?

   ○ Your endocrine system is trying to maintain homeostasis.
   ○ Your immune system is fighting the infection by cooking the pathogens.
   ○ You are dehydrated, which happens when you get sick.
   ○ Your heart is pumping faster to pump the pathogens out of your body

8. The organs of the immune system are connected by

   ○ lymph nodes.
   ○ Peyer's patches.
   ○ lymphatic vessels.
   ○ memory lymphocytes.

9. This organ has white blood cells that filter pathogens as they enter through your mouth and nose. They can become infected and are sometimes removed.

   ○ Spleen
   ○ Tonsils
   ○ Peyer's patches
   ○ Thymus

10. Where are memory lymphocytes made?

   ○ Lymph vessels
   ○ Peyer's patches
   ○ Thymus
   ○ Spleen

Fill in the Blanks

Use the word bank below to fill in the blanks in the story on the next page.

oxygen	spongy bone	nervous
cells	brain	muscles
immune and lymphatic	glucose	bladder
stomach	urinary	homeostatic feedback mechanism
muscular	digestive	circulatory
heart	blood	skin
nerves	endocrine	respiratory
white blood cells	lungs	skeletal

Pandia PRESS

Chapter 19: Show What You Know *continued*

You wake up and think about a dream you had last night. You wonder what part of your

_____ makes your dreams. Your _____ system receives a signal telling you to get up

because your _____ system just sent it the message that you have to pee. Your _____

must have filled up with urine while you were sleeping. While you are in the bathroom your

_____ starts to growl; your _____ system must be sending it signals, saying "Feed

ME!" Your _____ system is using the _____ _____ _____ to make

you hungry because your cells need _____ for cellular respiration so you can have

enough energy. You make a healthy breakfast because you know your _____ are going to

need lots of molecules to build more, because you are growing. In fact, you are probably

making more _____ right now in the _____ _____ part of your bones, so your

_____ system can carry food molecules throughout your body to those cells. You put

your hand in the middle of your chest and feel your _____ pumping that blood where it

needs to go.

It is raining. After breakfast you go outside and feel rain as it hits your _____. Your

integumentary system must have a lot of _____ running through it for you to feel rain hit

it. After catching some raindrops on your tongue, you breathe air deep into your _____

so your _____ system can have lots of _____ for all the puddle jumping you

have planned. Your _____ system gets ready as you tense your _____ so you

can jump from puddle to puddle. It's a good thing you have your _____ system to

support it all or you would be a puddle yourself. Oops, you slip and you cut yourself. You can

almost feel your _____ _____ _____ systems getting ready to do battle with

_____ _____ _____ rushing to the site.

 # Chapter 20: A Story of Luck

 # Evolution: A Timeline
Chapter 20: Lab

Have you ever noticed that when you study history, the closer you get to modern times the more information there is? Do you think this is because more happens now than happened in the past? Or do you think that when the ability to write became commonplace in civilizations, there was more information collected about that time? For example, we know much more about World War II than the Trojan War, and more about the Trojan War than the first recorded war between Sumer and Elam that was fought around 2700 BCE. There were most likely wars before 2700 BCE, but there are no records of them.

It is the same with the history of life on Earth. Earth is approximately 4.5 billion years old. There is more information about times closer to today than there is about 4.5 billion years ago. It does not mean that less happened; we just do not know exactly what it was. A good example of this is stromatolites. There are fossils of stromatolites that are 3.5 billion years old. The cell structure of stromatolites is too advanced for them to be the first organisms. What was before them, before 3.5 billion years ago? Scientists do not know, yet. Soft, unicellular organisms are tiny and decompose easily, especially after more than 3.5 billion years. This has made good, definitive evidence hard to find.

Keep this in mind when you make the timeline for this lab. The closer you get to today, the more types of organisms and the greater number of divisions can be put on the timeline. This is a reflection of our knowledge, not a reflection of diversity and change.

First, you will make a Geological Timeline. A Geological Timeline shows when the eons, eras, periods, and epochs started and finished. After you have made the Geological Timeline, you will add a Timeline of Evolution to it, showing the major evolutionary events that have occurred over the past 4.5 billion years.

Chapter 20: Lab *continued*

Timeline Vocabulary

The timeline is not divided at regular intervals. Scientists decide on the dates of the divisions based on major changes in organisms as recorded in rocks and in the fossil record.

Eon – An eon is the largest unit of time on the timeline. There will be 4 eons on your timeline.

Era – Eons are divided into eras. Each eon contains at least 2 eras. The most recent eon, the Phanerozoic Eon, is divided into 3 eras.

Period – Eras may be divided into periods.

Epoch – Periods may be divided into epochs. The most recent period, the Quaternary Period, is divided into 2 epochs.

For the purpose of this lab, you are given exact times for events instead of a range. For example, while Earth is estimated to be between 4.55 and 4.5 billion years old, you will be using the date of 4.5 billion years. All the events in the timeline happened. As to the exact date they happened, that is more speculative the farther away we are from today. ★At a certain point on your timeline (at the Phanerozoic Eon), you will be switching from "billions of years ago" to "millions of years ago."

Materials

- A workspace that is 45 feet long (preferred), or 45 inches—it all depends on your space constraints. This lab is fun to do outside, but can be done inside if that works best. The units will be scaled in the following way according to which unit you use: If you use 45 feet: 10 feet = 1 billion years
                                            If you use 45 inches: 10 inches = 1 billion years

- Tape measure that has the units you are using on it: feet or inches

- Marker

- Cardboard or cardstock paper. You will be cutting out the geological markers and gluing them to strips of cardboard or cardstock. Alternatively, you can copy the page onto cardstock and cut the cardstock into strips.

- 19 poker chips or checkers. You will be cutting out the evolution markers and gluing them to the chips or checkers.

- Glue

- Scissors

- Colored pencils to decorate the markers (optional)

- Roll of banner paper that is 45 inches or 45 feet long (if making your timeline into a banner)

Procedure

1. Decide where you are going to build your timeline. Are you going to make a timeline in your backyard or a park near you? Or are you going to make a banner that goes along a hall or around the walls in your classroom? Perhaps space is limited, and you need to make a timeline on your kitchen table that is 45 inches long.

2. Cut out all the markers from pages 283 and 285 (strips and circles) and glue them to strips of cardboard and poker chips. Color them if you wish. **Keep track of the number under the marker circles.**

Chapter 20: Lab *continued*

3. Make the Geological Timeline.

- Measure a line that is 45 units, feet or inches, long. (Or roll out a little more than 45 inches or feet of banner paper.)

- Mark one end "4.5 billion years ago." That is when Earth was formed. Mark the other end "today." That is the present day. This will be the longest measurement. All the other measurements are done inside or at the boundaries of this span.

- Put the eon markers on the timeline. An eon spans the time from when it began until when the next eon started. Mark each eon at its beginning. The first eon, the Hadean, began 4.5 billion years ago. All eon measurements are from the Hadean marker (4.5 billion years).

Eon	Beginning of Eon	Measurement and Marker Placement
Hadean	4.5 billion years ago	Put the marker at 4.5 billion years ago
Archean	3.8 billion years ago	Measure 7 ft (7 in) and put the Archean marker there
Proterozoic	2.5 billion years ago	Measure 20 ft (20 in) from Hadean and put the Proterozoic marker there
Phanerozoic	542 million years ago	Measure 39 ft 6 in (39.5 in) from Hadean and put the Phanerozoic marker there. The Phanerozoic Eon is the eon we live in now.

- Put the era markers on the timeline. Eons are made up of eras. You will only be recording the eras for our current eon (Phanerozoic). The first era of the Phanerozoic Eon started when the eon began, 542 million years ago. All era measures are from the Paleozoic marker.

Era	Beginning of Era	Measurement and Marker Placement
Paleozoic	542 million years ago	Put the Paleozoic marker at the start of the Phanerozoic Eon
Mesozoic	248 million years ago	Measure 2 ft 11 in (2.9 in) from Paleozoic and put the Mesozoic marker there
Cenozoic	65 million years ago	Measure 4 feet 9 in (4.7 in) from Paleozoic and put the Cenozoic marker there. This is our current era.

- Put the period markers on the timeline. Eras are made up of periods. You will record periods for the Mesozoic and Cenozoic Eras. The first period you will put on your timeline is Triassic. It began 248 million years ago, when the Mesozoic era began. All period measurements are from the Triassic marker.

Period	Beginning of Period	Measurement and Marker Placement
Triassic	248 million years ago	Put the Triassic marker at the start of the Mesozoic Era
Jurassic	206 million years ago	Measure 5 in (0.4 in) from Triassic and put the Jurassic marker there
Cretaceous	144 million years ago	Measure 13 in (1.1 in) from Triassic and put the Cretaceous marker there
Tertiary	65 million years ago	Put the Tertiary marker at the start of the Cenozoic Era

Chapter 20: Lab *continued*

Quaternary   2 million years ago   Measure 30 in (2.5 in) from Triassic and put the Quaternary marker there. This is the period we are in now.

- Put the epoch markers on the timeline. Periods are made up of epochs. You only record the epochs for the current period (Quaternary). The first epoch of the current period started when the period began 2.0 million years ago.

Epoch	Beginning of Epoch	Measurement and Marker Placement
Pleistocene	2 million years ago	Put the marker at the start of the Quaternary Period
Holocene	10 thousand years ago	The Holocene Epoch is 0.0012 inch (a smidgen) from now! This is our current epoch.

4. Now that you have made the Geological Timeline, it is time to add the Timeline of Evolution. Use the Timeline of Evolution lab sheet to fill in your timeline with the major milestones in evolution that have occurred. The table on the lab sheet tells you the marker number, the date and measurement to place the marker on your timeline, and the evolutionary event that occurred. Notice the change of "billions of years ago" to "millions of years ago" occurs between the advent of multicellular organisms and the first animals.

5. Find out which fossils have been found where you live and add a drawing or picture of it to the timeline. Google search for "Fossils in (the name of your state, country, or county)." Find out what your state fossil is, or use the site www.paleoportal.org to find fossils near you. Some areas, like many on the West Coast of the United States, will not have dinosaur fossils. Much of the West Coast of the United States was under water during the time of the dinosaurs. In that case, put your state's fossil in the timeline at the time it was alive and choose your favorite dinosaur to put in the timeline.

This timeline is incomplete. The lab would be much too long if all the known information was included in it. If you want to add more organisms, Google "When did _____ evolve?" Add the name, the date, and a picture of the organism to your timeline.

Optional Extras

1. Draw a picture of what your area looked like when the fossilized organism from your area was alive.

2. Write a story about a day in the life of this organism.

3. Visit an area where these fossils were found. If you can, lie on your back and imagine what the world was like then and all the changes that have happened to it since the time the fossilized organism lived.

4. Visit a museum in your area that has fossils.

Explore

Evolution: A Timeline

Chapter 20: Lab Sheet

Timeline of Evolution

★ The measurements for the first 5 markers are from 4.5 bya. Measurements starting with marker 6 (first animals) are from 1.5 bya.

Marker #	Date at marker (bya = billion years ago, mya = million years ago)	length to scale ft (in)	Plants	Animals	Eukaryotes	Prokaryotes
1	4.5 bya	Start		The formation of Earth		
2	3.5 bya	measure 10 ft (10 in) from 4.5 bya				1st prokaryotic cells, stromatolites
3	3.0 bya	measure 15 ft (15 in) from 4.5 bya				photosynthesis evolves
4	1.85 bya	measure 26 ft 6 in (26.5 in) from 4.5 bya			eukaryotic cell	
5	1.5 bya	measure 30 ft (30 in) from 4.5 bya			multicellular organisms	
Here there is a change from billion years ago to million years ago. The measures from here down are all from 1.5 bya (Marker 5).						
6	640 mya	measure 8 ft 7.2 in (8.6 in) from 1.5 bya		1st animals		
7	570 mya	measure 9 ft 3.6 in (9.3 in) from 1.5 bya		exoskeletons		
8	500 mya	measure 10 ft (10 in) from 1.5 bya		fish		
9	475 mya	measure 10 ft 3 in (10.25 in) from 1.5 bya	land plants			
10	400 mya	measure 11 ft (11 in) from 1.5 bya		insects		
11	360 mya	measure 11 ft 4.8 in (11.4 in) from 1.5 bya		amphibians		
12	300 mya	measure 12 ft (12 in) from 1.5 bya		reptiles		
13	290 mya	measure 12 ft 1.2 in (12.1 in) from 1.5 bya	gymnosperms			
14	230 mya	measure 12 ft 8.4 in (12.7 in) from 1.5 bya		dinosaurs		
15	200 mya	measure 13 ft (13 in) from 1.5 bya		mammals		
16	150 mya	measure 13 ft 6 in (13.5 in) from 1.5 bya		birds		
17	140 mya	measure 13 ft 7.2 in (13.6 in) from 1.5 bya	angiosperms			
18	65 mya	measure 14 ft 4.2 in (14.35 in) from 1.5 bya		dinosaurs go extinct		
19	200,000 years ago	place near the end of the Pleistocene epoch		humans		

Chapter 20: Geological Markers

Eon Markers

Hadean Eon started 4.5 billion years ago
Archean Eon started 3.8 billion years ago
Proterozoic Eon started 2.5 billion years ago
Phanerozoic Eon started 542 million years ago

Era Markers

Paleozoic Era started 542 million years ago
Mesozoic Era started 248 million years ago
Cenozoic Era started 65 million years ago

Period Markers

Triassic Period started 248 million years ago
Jurassic Period started 206 million years ago
Cretaceous Period started 144 million years ago
Tertiary Period started 65 million years ago
Quaternary Period started 2 million years ago

Epoch Markers

Pleistocene Epoch started 2 million years ago
Holocene Epoch started 10 thousand years ago

Chapter 20: Evolution Markers

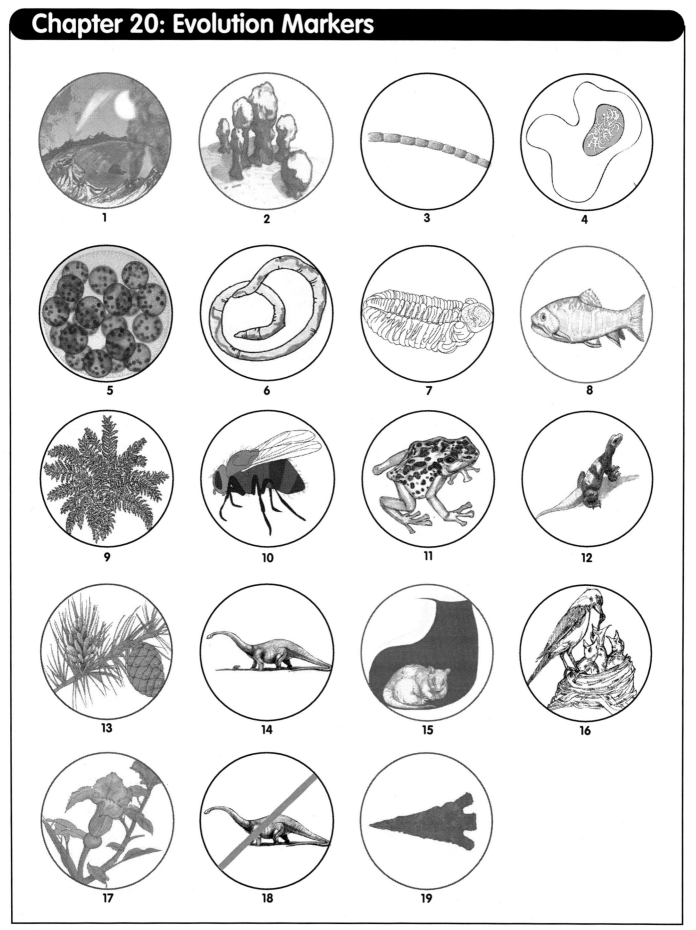

On the Wings of a Bug

Chapter 20: Microscope Lab

Scientists believe that all insects evolved from a common ancestor over 400 million years ago. The **common ancestor** of a group of organisms is the species from which all the organisms in the group evolved. Therefore, the species that gave rise to all insects lived about 400 million years ago, and it is the common ancestor to all insects.

The earliest insects were small and wingless. The first fossils of winged insects are 350 million years old. Over the past 400 million years, insects have evolved into the millions of different species of insects that have lived since then. Today you will examine the wings of at least two different species of insects with your microscope.

Materials

- Microscope
- Slide
- Two or more different species of dead winged insects; more is better

- A flashlight or desk lamp, in case you need top lighting
- Scalpel or paring knife

Procedure

If the insects are old, they might be brittle. Be very careful, or you can get insect parts in your microscope.

1. Collect the wings of different species of insects.

2. Cut off a wing of one of the insects.

3. Make a dry mount slide with the wing.

4. Examine it with your microscope.

5. Draw the microscope view on your lab sheet.

6. Repeat the procedure for all the insect wings you have collected.

7. If you are having trouble seeing the sample, try looking at it with top lighting.

8. On the lab sheet, record the name of each insect, if you know it. If you can't discover the name of the insect, then describe it. Compare the similarities and differences between the wings.

Pandia PRESS

On the Wings of a Bug

Chapter 20: Microscope Lab Sheet

Name_____ **Date**_____

Type of mount _____ Microscope_____

Microscope View	Name/Description of Insect	Comments: Compare/Contrast

Pandia PRESS

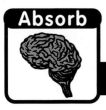

The Burgess Shale

Chapter 20: Famous Science Series

One of the Burgess Shale organisms

Famous Fossil Discoveries

In 1909 Charles Walcott, a paleontologist with the Smithsonian Institute, was riding his horse in the Canadian Rockies. A **_paleontologist_** is a scientist who studies ancient life by collecting and analyzing fossils. One of the mules in the group lost a shoe. When Walcott dismounted, he saw a rock and picked it up. Being a paleontologist, he cracked the rock open to see what was inside it. What did Charles Walcott see?

Where is the Burgess Shale site? It is named after Mount Burgess, where the fossils are found. Locate Mount Burgess on a map.

Chapter 20: Famous Science Series *continued*

How many types of fossilized organisms have been found at this site?

List at least two things that make the Burgess fossils special.

What happened to the organisms over 500 million years ago that created the Burgess Shale site?

Pandia PRESS

A Story of Luck

Chapter 20: Show What You Know

Multiple Choice

1. Life on Earth evolved

   ○ 450 million years ago.
   ○ 10 billion years ago.
   ○ before 3.5 billion years ago.
   ○ after 450 million years ago.

2. Three billion years ago, photosynthesizing organisms evolved. They spewed a waste product into the water and air. What was the waste product?

   ○ Carbon dioxide
   ○ Oxygen
   ○ Ozone
   ○ Nitrogen

3. From question 2: The evidence for this is found all over the world in

   ○ banded iron formations.
   ○ volcanoes.
   ○ the ocean.
   ○ metamorphic rocks.

4. Biological evolution

   ○ happens to an individual.
   ○ happens to a group.
   ○ never happens.
   ○ always results in a new species.

5. The oldest fossils are _____. They are_____.

   ○ dinosaurs, 450 million years old
   ○ eukaryotic cells, 1.85 billion years old
   ○ plants, 450 million years old
   ○ stromatolites, 3.5 billion years old

6. An example of a population is

   ○ all the penguins on Earth today.
   ○ all the different species of penguins in all the areas of Antarctica that can breed.
   ○ all the Emperor penguins in all the areas of Antarctica that can breed.
   ○ all the Emperor penguins that live and breed together in a small area in the Weddell Sea region of Antarctica.

7. The Endosymbiotic Theory explains how

   ○ plants evolved.
   ○ animals evolved.
   ○ multicellular organisms evolved.
   ○ eukaryotic cells evolved.

8. Geologic time

   ○ is the time it takes rock to form.
   ○ is the time from when the earth was formed to the present day.
   ○ is used by geologists to time geological processes.
   ○ is how long it takes mass extinctions to occur.

Chapter 20: Show What You Know *continued*

9. The creation of an ozone layer was important because

   - ○ it created a protective shield from the sun's rays so that organisms could colonize land.
   - ○ it made photosynthesis possible.
   - ○ oxygen is necessary for cellular respiration.
   - ○ it traps warm light from the sun, which makes the earth warm enough to support life.

10. A group of organisms called archosaurs gave rise to

    - ○ birds.
    - ○ dinosaurs.
    - ○ crocodiles.
    - ○ All of the above

11. What trait did plants and animals have at the time that made the colonization of land possible?

    - ○ Mitochondria
    - ○ Eukaryotic cells
    - ○ A protective outer layer
    - ○ They were multicellular

12. Mass extinctions occur when

    - ○ all the populations of a species die.
    - ○ organisms evolve.
    - ○ a large number of species die within a short period of geologic time.
    - ○ all of one type of organism dies, like when the dinosaurs died.

Questions

1. Dinosaurs roamed the earth for about 1.65 million years. Sixty-five million years ago an asteroid hit Earth at the same time that massive volcanic eruptions were occurring. Explain how these two catastrophes could have led to the mass extinction of the dinosaurs.

Chapter 20: Show What You Know *continued*

2. There were several evolutionary steps going from simple prokaryotic organisms to complex multicellular eukaryotic organisms. Give a step-by-step explanation for how this could have occurred.

3. When dinosaurs went extinct, another group of animals, mammals, possessed a number of traits that made it possible for them to survive the environmental catastrophes occurring on Earth. What were these traits, and how did they benefit mammals?

Chapter 21: How

Explore

Natural Selection

Chapter 21: Lab

If you were a mouse, what color would you want to be?

Have you ever been to a pet store and observed all the different colors of mice they have? Have you ever wondered why wild mice are brown or brownish-gray? Why aren't they black, white, or spotted like pet store mice? Think about mice in all these colors running across the lawn. Now imagine a hungry owl is sitting in a tree. The owl hears all the mice. He looks around. Which mouse will be his dinner? Mice are quick. When the owl flies down from the tree, he has one chance to catch dinner before the mice can get back to their burrow. Which mouse do you think the owl will see first? Which mouse will stand out the most on the surface? Which mouse is the most likely to be caught and eaten?

This is how selection works. The mouse that is caught and eaten will not produce any more offspring. It will not pass any more of its genes on to future generations. The mice that are most likely to escape are those that are best adapted to their environment. Which mice have the best color adaptation for this environment: the black, white, spotted, brown, or brownish-gray mouse?

Today you are going to be a hungry owl. You will use pompoms for the mice. A helper will scatter the "mice." You will have just seconds to "catch" all the mice you can. Which color mice do you think you will catch the most of, and which mice are best adapted to their environment?

Pandia PRESS

Chapter 21: Lab *continued*

This lab can be done outside or inside. The list of materials is slightly different depending on where you conduct it. If you do this lab outside, the weather can affect the outcome. During times of snow, some mice burrow through the snow instead of spending time on top of it. Why do you think that is?

Materials

Outside List

- Pompoms in each of the colors listed below. You need at least one color pompom that really blends in with the surface where they are spread. For example, add 20 light green pompoms if your outside area is grassy. There will be pompoms that are not used, but you cannot assume which color this will be.

    20 brown pompoms

    20 black pompoms

    20 gray pompoms

    40 white pompoms

    20 additional pompoms if your chosen surface is not brown, black, or gray

- Black magic marker

- An extra person to scatter the "mice"

- Timer

Inside List

- All the materials from the outside list. (The 20 additional pompoms should blend into the color of your inside surface. For example, if you have violet shag carpet (nice!), you will need 20 violet pompoms.)

- 2- to 3-meter surface to put the pompoms on that has some depth to it. Shag carpet, furry fleece, faux bear fur, or other rough cloth will work.

Procedure

1. Using the black marker, put black spots on 20 of the white pompoms.

2. Write your hypothesis on the lab sheet. Will there be selection for the "mice"? If yes, which color do you think will be selected most strongly for, and which will be selected most strongly against when you try to catch them?

3. Have another person scatter **five** of each color of pompoms in a 2- to 3-meter square area. Do not peek to see where they are. Close your eyes until they lead you to one side of the area.

Chapter 21: Lab *continued*

4. **Note to the person scattering the pompoms:** Scatter the pompoms randomly. Make sure not to bunch one color all together. Try to cover the area. If possible, put the pompoms slightly down into the nap of the surface material. The person scattering the mice must time the "owl." When you are done scattering the mice, lead the "owl" person to the area. Make sure the "owl" does not observe the area until you are ready with the timer. Count 1, 2, 3, go, and start the timer for 30 seconds while the "owl" catches mice. At the end of 30 seconds, say *stop*.

5. **Do not peek until the scatterer says go.** On the count of 1, 2, 3, go! Open your eyes, survey the landscape, and start catching mice. Toss the first mouse aside so you can remember what mouse that was when the 30 seconds are done. When the person timing says stop, you need to immediately stop collecting mice.

6. Pick up the first mouse you caught and record the color. Record the rest of the pompoms you caught on the lab sheet. Subtract the number of mice caught from the number that were scattered when you started the round. Put the pompoms you caught aside. They are done.

7. The "mice" you did not catch have now reproduced and passed on their fur color genes to their offspring. Add one mouse to the area for every mouse that was not caught, matching the colors. The number of mice that were not caught, going into the next round, has now doubled.

8. **Round two:** Repeat the above procedure from #3 through #7.

9. Record your results on the lab sheet.

10. **Round three:** Repeat the above procedure from #3 through #7.

11. Finish filling in your lab sheet.

12. Complete a formal Lab Report for this lab.

Explore

Natural Selection
Chapter 21: Lab Sheet

Name_____ Date_____

Objective: To determine which mice have the best color adaptation for their environment.

Hypothesis:

Results and Observations

Round 1 – color of first mouse caught:

Round 2 – color of first mouse caught:

Round 3 – color of first mouse caught:

Data Table

Color / Round		Black	Brown	Gray	Spotted	White	
1	Amount caught						
	Amount left						
	Amount added						
	Total mice for next round						
2	Amount caught						
	Amount left						
	Amount added						
	Total mice for last round						
3	Amount caught						
	Amount left						

Pandia PRESS

Chapter 21: Lab Sheet *continued*

Questions

Did you catch the same color mouse first every time? What do you interpret from that?

Which colored mouse was best adapted for the environment? Which colored mouse was the worst adapted for the environment?

Based on the results from this experiment, why do you think the mice in the wild are brown or brownish-gray in color? Use the term *natural selection* in your answer.

In terms of evolution, fitness is defined as the ability to produce offspring. Which fur color results in the best fitness for the mice?

If there is continued selection for and against certain fur colors, what do you think will be the color of the mice in this population?

What happened to the mouse that had the best fur color once it became more numerous?

Natural Selection

Chapter 21: Lab Report

Name_____ Date_____

Title/Location_____

Hypothesis

Procedure

Observations

Results and Calculations

Conclusions

Function and Form

Chapter 21: Microscope Lab

Fur, hair, wool—call it what you want, but scientifically it's all fur. All mammals have it. The first definitive fossil with fur is an animal from 164 million years ago. Fur doesn't fossilize well, so no one can be sure when it first evolved. The function of fur is to insulate and protect. Why are there so many different forms, though, in all those different colors? The evolution of fur between different species of mammals, and even within a species, has led to many different forms, but it all has the same function. Today you will look at some different forms of fur under the microscope.

Materials

- Samples of hair, fur, and/or wool from as many different types of mammals as possible (dog, cat, hamster, sheep, cow, human, chinchilla, horse, etc.)
- Microscope

- Slides (as many as your fur samples)
- Slide covers
- Water
- Dropper

Procedure

1. Answer the initial questions on the lab sheet.

2. Make wet mount slides one at a time, one for each fur type. Do not clean slides until the end, in case you want to refer back to a slide.

3. Look at them under the microscope.

4. Take notes on the lab sheet as you go along.

Function and Form

Chapter 21: Microscope Lab Sheet

Name_____ Date_____

Before You Begin

All these questions refer to the texture of the fur, not the color.

What two fur samples from different types of animals do you think will look the most alike with a microscope? Why?

What two fur samples from different types of animals do you think will look the most different with a microscope? Why?

While You're Working

List the name of each type of fur (name of the mammal it came from). Jot down notes about the appearance of each. Compare the fur samples with each other. Write down any differences you observe.

Type of Fur	Appearance Description	Comments: Compare/Contrast

Pandia PRESS

Chapter 21: Microscope Lab Sheet *continued*

Type of Fur	Appearance Description	Comments: Compare/Contrast

Conclusions

What two fur types looked the most alike with a microscope? Why?

What two fur types looked the most different with a microscope? Why?

Were you surprised by anything you saw? If so, what?

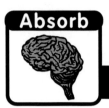

Evolution Act 1: First Theories
Chapter 21: Famous Science Series

Jean-Baptiste Lamarck

Imagine you live your whole life with no car, no bus, no train, and no plane. Some people have horses and buggies, but not many. Imagine that you walk everywhere you go. There are no televisions, there are no movies to watch, and no video games to play. You spend your entire life in a small area, looking at the plants and animals right where you live. You might see an exotic plant or animal, but only two or three times in a lifetime, if that many.

This would have been what it was like for many western Europeans in the fifteenth century. By the late 1400s onward, explorers returned to Europe with fantastic stories of animals, plants, and people they had never seen before. As people began traveling the globe, there came an understanding that the world was filled with all sorts of different organisms. Another surprise was that the plants and animals you looked at your whole life were not in other places around the world. There were no horses on the American continents. In Australia, there were no rabbits or dogs, but there were koala bears and duck-billed platypuses. In China, there were funny-colored black and white bears. In the Arctic, the bears were white. The insects and plants were different. People in Europe were amazed and intrigued. In England in the 1800s there was even a beetle-collecting craze. Some people had large collections of beetles from locations around the globe.

People started to wonder, why were there different organisms in different locations? Why were there so many different types when fewer would work just as well? Then there was the fossil evidence. People had been discovering fossils for a long time. People wondered what had happened to all the types of organisms that had gone before and no longer existed.

Chapter 21: Famous Science Series *continued*

In 1809, Jean-Baptiste Lamarck proposed a theory of evolution. What was his theory and was it correct?

In 1788, James Hutton, a Scottish farmer and geologist, put forth a theory called uniformitarianism. What was the theory and was it correct?

Evolution: How
Chapter 21: Show What You Know

Multiple Choice

1. One bacterium splits into two, then two to four... Soon there are millions. The bacteria run out of food and begin to starve to death. This is an example of:

   ○ Natural selection
   ○ Extinction
   ○ Overproduction
   ○ Genetic drift
   ○ Speciation

2. The case of the peppered moths is a good example of how _____ works.

   ○ natural selection
   ○ reproductive isolation
   ○ mutation
   ○ speciation

3. What two mechanisms lead to genetic variation?

   ○ Overproduction and natural selection
   ○ Genetic recombination and mutation
   ○ Natural selection and reproductive isolation
   ○ Speciation and adaptation

4. Aquatic birds, like ducks, have webbed feet that help them paddle through water. This is an example of

   ○ genetic drift.
   ○ natural selection.
   ○ genetic variation.
   ○ an adaptation.

5. Dogs and cats are not the same species because

   ○ cats are carnivores and dogs sometimes eat grass.
   ○ they have reproductive isolation.
   ○ they cannot breed with each other and have offspring.
   ○ they have the same number of chromosomes but different genes.

6. Which of the following statements are true?

   ○ Mutations can be passed from parent to offspring.
   ○ Mutations lead to genetic variability.
   ○ Mutations can cause traits that are beneficial, neutral, or harmful.
   ○ All of the above

Chapter 21: Show What You Know *continued*

7. A population of beetles lives on an island. The beetles come in two colors green and brown. A hurricane blows five green beetles off their island onto another where there are no beetles. It does not take long for the beetles to colonize and have a healthy beetle population on their new island. All the beetles on the new island are green. This change to an all-green population on the new island is an example of

   ○ genetic drift.

   ○ genetic recombination.

   ○ genetic variation.

   ○ speciation.

8. Arctic hares have fur that is brown in the summer and white in the winter. If the earth became warmer and all the snow melted in the Arctic, this would be an example of a _____ trait that became a _____ trait.

   ○ neutral, harmful

   ○ beneficial, harmful

   ○ beneficial, neutral

   ○ harmful, beneficial

9. Reproductive isolation is necessary for speciation because

   ○ harmful mutations can accumulate and extinction will occur without it.

   ○ different species cannot reproduce with each other.

   ○ gene flow must be stopped between populations for one to evolve into a new species.

   ○ mutations are more likely to occur when organisms are isolated.

10. Mutations are _____; selection for the traits they cause is not.

    ○ common

    ○ uncommon

    ○ random

    ○ dangerous

11. What is the name of the process that explains how all the species of organisms have come to be?

    ○ Evolution

    ○ Genetic recombination

    ○ Genetic drift

    ○ Meiosis

12. Overproduction should lead to there being many more organisms alive than the earth can support. What are the controls on overproduction?

    ○ Predation

    ○ Disease

    ○ Scarce food resources

    ○ Weather

    ○ All of the above

Chapter 21: Show What You Know *continued*

Essay and Timeline

1. The sealocrab and the aquanotic are two different species of marine mammals that were discovered on the Island of Mythical Creatures. The sealocrab is the ancestral species to the aquanotic. On a separate sheet of paper, write a story of a possible scenario for how the evolution from sealocrab to aquanotic could have happened. Use some or all of the following terms:

genetic variation	genetic recombination	mutation	reproductive isolation
speciation	natural selection	genetic drift	adaptation
time	traits	evolve	evolution

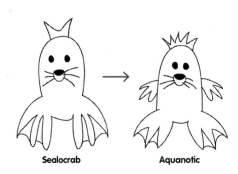

Sealocrab   Aquanotic

2. You learned about the cell theory in Chapter 2. Along with the theory of evolution, this is one of the central theories of biology. Below is a timeline in the development of the cell theory. At what point did the cell theory actually become a theory? To answer this question, use the definition and explanation of what scientific theory is. Circle the point on the timeline where the cell theory becomes a theory. If you are working on a history timeline, add some or all of the cell theory development dates to your timeline.

   1665 Robert Hooke discovers squarish-looking structures in a specimen of cork with his microscope. He names these structures *cells*.

   1674 Antonie van Leeuwenhoek is the first person to see a live cell with his microscope.

   1831 Robert Brown discovers the cell nucleus.

   1839 Thoedor Schwann and Matthias Schleiden propose a theory stating that all living things are made of one or more cells. Schwann and Schleiden perform many different experiments before proposing their theory. They conduct many experiments after their theory is proposed.

   1855 Rudolf Virchow proposes that every cell comes from another cell. This is added as a part of the cell theory. He performed many different experiments before proposing this. He conducts many experiments after his theory is proposed.

   1855 to present day. The cell theory has been tested many times by many different researchers. Their results have confirmed the cell theory.

Chapter 22: Evidence

Explore

Forming Fossils

Chapter 22: Lab

This fossil bed shows trace fossils created by the organisms that lived in them. It comes from Chicken Corner in Utah.

Organisms are born, they live their life, and then they die. With all this talk about fossils, you might think most of the organisms that die end up as fossils. Actually, fossils are not common. Most organisms are eaten or decompose without leaving a fossil. In almost all cases, it is the hard parts of organisms, like bones and shells, that fossilize. There is almost no fossil record of soft-bodied organisms like those found in the Burgess Shale. The most likely scenario for the fossilization of an organism is if the organism dies in a watery area and is buried with sediments soon after it dies.

Fossilized remains are in the form of body fossils or trace fossils. **Body fossils** are the preserved remains of all or some parts of an organism's body. Trace fossils are evidence left behind by an organism, like a footprint. Body fossils fossilize either with alteration (change) or without alteration. The preservation of fossils with alteration is a chemical process. The cells and tissues that were part of the organism's body chemically change into a different material.

One method of fossilization with alteration is **permineralization**. With permineralization, the sediment itself changes into rock, and so does the part of the organism's body that fossilizes.

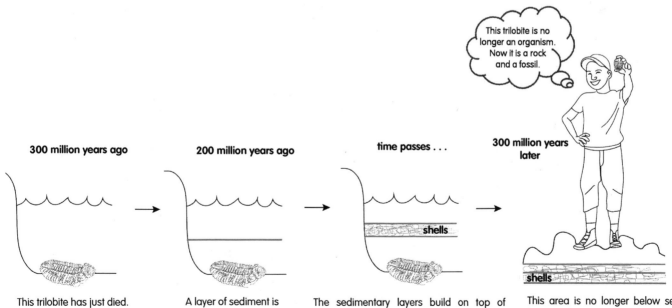

This trilobite is no longer an organism. Now it is a rock and a fossil.

300 million years ago

200 million years ago

time passes . . .

shells

300 million years later

shells

This trilobite has just died.

A layer of sediment is covering the trilobite.

The sedimentary layers build on top of the trilobite. As this happens, minerals in the water that are in the sediment replace the organic minerals that were part of the trilobite's cells. This is called permineralization. Permineralization creates a rock-shaped copy of the trilobite.

This area is no longer below sea level. As the rock wears away from erosion, this trilobite will once again see the light of day as a fossil.

Chapter 22: Lab *continued*

Carbonization is another type of fossilization with alteration. Carbonization occurs through a chemical reaction where the plant or animal is transformed into a thin film of carbon.

The preservation of fossils without alteration means that the organism, or most often parts of the organism, becomes a fossil without the organic material changing into another type of material. Fossils preserved without alteration are usually shells, teeth, and bones.

Have you ever seen an animal burrow, footprint, bird's nest, or scat (a nice name for poop)? These are evidence of animals living in that environment. If these are preserved, they become trace fossils.

Sometimes the original remains of the organism are gone and all that is left is an organism shaped hole in the rock. This type of fossil is called a *mold*. The mold can become filled in with sediments, and this is called a *cast fossil*. A cast fossil is a three-dimensional representation of the original. Molds and casts show what an organism or part of an organism looked like, but they are not its fossilized remains.

Today you will make a mold into clay with various objects. You will carefully press each object into the clay, and then just as carefully remove them. The imprints left behind are called molds. After that, you will pour Plaster of Paris over the clay. The Plaster of Paris will act as sediments do in authentic cast fossils. It will fill the molds. You will have created a cast fossil that is a replica of a fossil bed.

This footprint cast fossil was made by an allosaurus in Moab, Utah, during the Jurassic period 200 to 145 million years ago.

A thin film of carbon in the shape of a leaf is all that remains of the fern leaf that lived millions of years ago. This is an example of carbonization.

A fossil without alteration is this baby woolly mammoth. In 1977, this 7- to 8-month-old was discovered frozen in the permafrost in Siberia. The frozen climate kept much of its body preserved. Radioactive carbon dating determined the mammoth died about 40,000 years ago.

Chapter 22: Lab *continued*

Materials

- Aluminum or silicon pie or cake pan
- Clay to press into the pan two inches thick, modeling clay works best
- Cooking spray oil
- Items to create fossils with: shells, bones, sturdy leaves (like fern), rocks, sticks
- 3 cups Plaster of Paris
- Container for mixing Plaster of Paris
- Something with which to stir Plaster of Paris
- Measuring cup
- 1½ cups water
- Towel

Procedure

1. Cover the bottom of the pan with 1 to 2 inches of clay. Make sure it is very smooth.

2. Press the items to be fossilized into the clay. DO NOT wiggle them when you pull them out. Pull them straight out.

3. Spray oil over the sides of the pan, the clay, and impressions.

4. Make the Plaster of Paris. Mix 3 cups of Plaster of Paris powder with 1½ cups of water. Let this sit for 1 minute. Do not let this sit longer than that; Plaster of Paris hardens very quickly.

5. Pour the Plaster of Paris over the clay. Fill the impressions in first. Then fill the rest of the pan so that the Plaster of Paris is 1 to 2 inches thick.

6. Let the Plaster of Paris harden. This will take 2 to 4 hours.

7. Put the towel over the top of the pan. Turn the pan over with your hand on it. Put the pan down, towel side down. Lightly tap the pan so the Plaster of Paris mold separates from the clay. This is your fossil bed.

8. Store the clay in an airtight container for future use.

Layers

Chapter 22: Microscope Lab

Each layer in this sedimentary rock was laid down, one on top of the other. The oldest layer is at the bottom and the newest layer is at the top.

Most fossils are found in sedimentary rock. Sedimentary rock is made of sediment that settles in layers. It is exposed to pressure, and the layers of sediment become rock. The remains of organisms that die can be trapped within the layers, becoming fossils. When you look at sedimentary rock, you can see the layers. If you look at the rock with your microscope, you can see the individual pieces of sediment. Do you think you can see the layers with your microscope? Maybe you will even see a fossil.

Materials

- Microscope
- Light for top lighting—another person holding a flashlight works best
- Sedimentary rock (not too big; it must fit under your microscope). Sandstone, limestone, conglomerate, and shale are sedimentary.
- Paper to put the rock on, so no crumbs fall down into the bottom of the microscope
- Pencil with an eraser
- Optional: magnifying glass

Procedure

1. Examine the rock without the aid of the microscope. Draw a picture of it on your lab sheet.

2. Put a sheet of paper on the stage of the microscope. This is so no grains from the rock fall down into the bottom of the microscope.

3. Put the rock on the paper. Make sure it fits under the microscope lens! The rock will damage the lens if it scrapes against it.

4. Set up the light so it will shine on the rock.

5. Look at the rock on power 40x. Do not move the lens. This is the strongest power you are going to use to observe the rock.

6. Draw the view of the rock as seen through the lens. Look at the rock from all sides. If you have trouble getting a good view with your microscope, look at the rock with the magnifying glass.

Layers

Chapter 22: Microscope View Sheet

Name_____ **Date**_____

Specimen _____ Type of mount_____

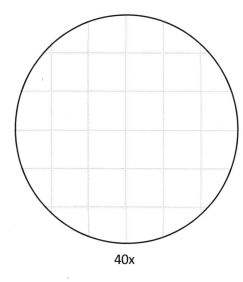

40x

Do you see the divisions between layers?

Different-colored rocks have different chemical compositions. This means conditions in the location were different when these different-colored layers were put down. Do you see rock layers that have different chemical compositions from one another?

Do you see any fossils in your sedimentary rock?

Additional comments:

Absorb

Evolution Act 2: Charles Darwin
Chapter 22: Famous Science Series

In 1831, Charles Darwin began a journey circumnavigating the globe by sea. Darwin was a thoughtful observer of nature. He spent a lot of his life puzzling over things he saw. Early in the voyage, he stopped at the Cape Verde Islands in Africa. There he saw, 45 feet above sea level, a white stripe of fossilized shells running through the rocks around the islands. Darwin wondered how these shells came to be so far above the sea. He asked himself: Why were there different organisms in different locations? Why were there were so many different types when fewer would work just as well? Then there was the fossil evidence. People had been discovering fossils for a long time. Darwin wondered what had happened to all the types of organisms that had gone before and no longer existed. Darwin had read about the theory of uniformitarianism and he wondered that if the earth was constantly changing, was it possible that organisms had been changing too?

Look up Darwin's voyage that began in 1831. Draw his path on the map below.

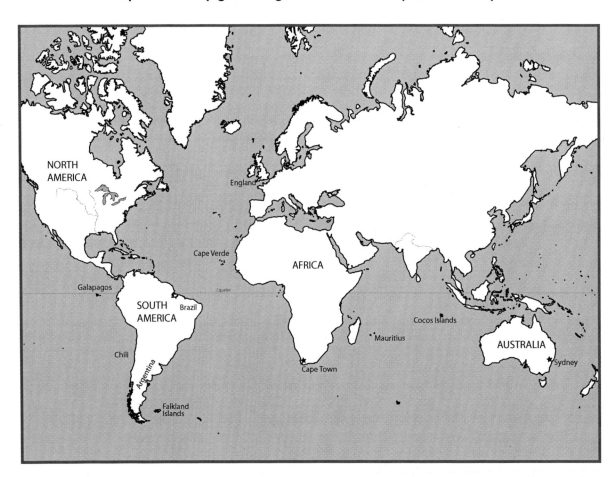

Chapter 22: Famous Science Series *continued*

What was the name of the ship Darwin sailed on?

How long was he at sea?

What was his job on the ship?

Darwin spent only 18 months of the journey at sea. The rest of the time he spent on land collecting plant, animal, and fossil specimens and making geological observations. Some of the most important collecting was done during the five weeks he spent on the Galápagos Islands. The Galápagos Island chain is 600 miles west of the west coast of South America, in the Pacific Ocean. Different islands in the chain have different environments, with different types of plants and insects growing on them. Most of the islands are far enough from each other so that there is reproductive isolation for many of the species on each island. Darwin visited four of the over 14 islands in the chain. While on the islands, he collected finches.

Darwin returned to England with 13 different species of finches, each with its own beak shape and size, collected on the Galápagos Islands. This started Darwin wondering why each island had a different finch species. There was only one species of finch on the South American mainland. He puzzled over why there were 13 species only 600 miles away. What did he conclude was the answer?

The finches from the Galápagos Islands are one of the many examples used in Darwin's book showing evidence of evolution. What was the complete title of his book? (It's 21 words long!) What is the shortened title most people use?

How long did it take Darwin to write it?

Why did he finally publish it?

There was one weakness in Darwin's book. Darwin did not provide the mechanism for how traits are passed on from parents to their offspring. That is because Darwin did not know it.

Evolution: Evidence
Chapter 22: Show What You Know

Multiple choice

1. A sequence of fossils have been found that show the evolution of horses from fox-sized, forest-dwelling animals with four toes on its front feet and three on its back to the large, one-toed animals we have today. This is an example of

   ○ homology.
   ○ transitional fossils.
   ○ biogeography.
   ○ artificial selection.

2. In what kind of rocks do fossils form?

   ○ Igneous
   ○ Metamorphic
   ○ Sedimentary
   ○ Volcanic

3. The more closely related two species are, the more homologous their biology and chemistry is. In this sentence the word *homologous* means

   ○ similar.
   ○ dissimilar.
   ○ advanced.
   ○ simple.

4. Body parts that share a common ancestor but serve a different function are called

   ○ vestigial structures.
   ○ convergent structures.
   ○ divergent structures.
   ○ homologous structures.

5. What body parts are the most likely to become fossils?

   ○ Soft body parts
   ○ All parts are equally as likely to fossilize
   ○ The hard parts like shells, wood, and bones
   ○ None—that is why there are cast and mold fossils

6. What molecule have scientists recovered from some insects trapped in amber and frozen woolly mammoths?

   ○ Carbohydrates
   ○ DNA
   ○ Proteins
   ○ Lipids

Chapter 22: Show What You Know *continued*

7. The bubonic plague is a deadly infection caused by the *Yersinia pestis* bacteria. The sickness it causes is sometimes called the Black Death. It is believed the bubonic plague killed millions of people in Europe, one-third of the overall population, over the course of 400 years. Some people who lived in plague-hit areas survived and were resistant to contracting the disease. They passed their resistance on to their offspring. This is an example of

   ○ artificial selection.
   ○ homology.
   ○ natural selection.
   ○ biogeography.

8. Humans have intentionally bred milk cows for increased milk production. What is this intentional breeding called?

   ○ Natural selection
   ○ Artificial selection
   ○ Speciation
   ○ Genetic variation

9. Species have traces of DNA from distant ancestors. Sometimes this DNA codes for structures a species does not use, structures that are homologous to functional structures in other species. These are called

   ○ divergent structures.
   ○ convergent structures.
   ○ homologous structures.
   ○ vestigial structures.

10. For many ground-dwelling organisms, the Grand Canyon is a deep, wide, impassable expanse. Rodent species are different from one side of the Grand Canyon to the other, while bird species are the same on both sides. This distribution of animal species at the Grand Canyon is explained by

    ○ biogeography.
    ○ genetic variation.
    ○ erosion.
    ○ continental drift.

11. With permineralization,

    ○ the organism is trapped in resin, which turns into the mineral amber.
    ○ a three-dimensional representation of the organism is fossilized as minerals fill in the space where the organism's body was.
    ○ an organism's soft tissue is replaced with minerals and creates a fossilized organism that is a rock.
    ○ an organism turns into a mineralized carbon trace.

12. An example of an organism that fossilizes without alteration is

    ○ a three-dimensional representation of an organism that is fossilized in the space the organism's body was.
    ○ an organism's soft tissue is replaced with minerals and creates a fossilized organism that is a rock.
    ○ a mammoth that is frozen in the tundra for 24,500 years.
    ○ an organism turning into a carbon trace.

Chapter 22: Show What You Know *continued*

13. A carbon imprint found on a rock is an example of

    ○ a chemical change.
    ○ fossilization with alteration.
    ○ carbonization.
    ○ All of the above

Questions

Why do fossils form in sedimentary rock and not the other types of rock?

The main evidence of evolution comes from four sources. What are they? Give a brief explanation of how each is evidence that evolution occurred.

Matching. Match the homology on the left with the example for that homology on the right.

Chemistry ○   ○ Pythons have tiny, useless hind legs, called spurs. It is believed these are left over from before snakes evolved from lizards.

Genetics ○   ○ Brown bats and manatees have similar bones in their arms.

Embryos ○   ○ A mouse and a mushroom are both made from hydrogen, oxygen, carbon, nitrogen, calcium, and potassium.

Vestigial Structures ○   ○ An ostrich and an owl have more gene sequences in common with each other than they do with a trout.

Homologous Structures ○   ○ The embryo of a raven and a rabbit look very similar.

Chapter 23: When

Are We Just Relatively Dating?

Chapter 23: Lab

The lab for today is a relative dating puzzle. The object of the puzzle is to organize groups of fossils into the time periods when they lived. The fossils are drawings of actual fossilized organisms that were found in layers of sedimentary rock in the Grand Canyon. Just like in nature, the oldest fossils were found in the bottom layers from more distant time periods, and the newest fossils were in the top layers from more recent time periods. You will have to use what you have learned in this unit to put the strips in the correct order.

Materials

- Scissors
- Glue
- Fossil Puzzle Cutouts sheet

- Geological Time Period sheet
- Large sheet of construction paper (optional)

- Internet connection is helpful to answer questions

Procedure

1. You need two sheets for this puzzle: The Fossil Puzzle Cutouts sheet shows groupings of organisms. These are your puzzle pieces. The second sheet has geological time periods. Each strip from the puzzle sheet has a corresponding time period it matches on the Geological Time Period sheet.

2. Cut out the puzzle pieces on the Fossil Puzzle Cutouts sheet.

3. Think about the order the pieces go on the Geological Time Period sheet based on what you know about the Principle of Superposition and everything you know about how life evolved. The oldest geological period is at the bottom and the most recent is at the top of the Geological Time Period sheet.

4. You can glue the fossil strips on the Geological Time Period when you are sure you have them in the correct order, or you can attach the Geological Time Period onto a large sheet of construction paper and glue them side by side onto that. BEFORE YOU GLUE: Show your work to your parent or teacher to check if you are correct. When you are certain, glue the pieces in the correct place.

Pandia PRESS

Chapter 23: Lab *continued*

5. Below are a few hints. First, try to see if you can put the organisms into the Geological Time Period without looking at the hints.

6. Answer the questions on the lab sheet.

Hints

Look at the hints one at a time. Maybe you will not need them all to solve the puzzle.

- If you are having trouble placing one of the strips, look at the strips you have put above and below as a guide.

- The first fossils are simple, unicellular organisms.

- The fossils show that organisms became increasingly more complex.

- Most of the time (but not always) there are more species of different organisms in the most recent layers.

- Organisms with an exoskeleton, an external skeleton, came before those with an endoskeleton, an internal skeleton.

- Marine organisms evolved first.

- Reptiles evolved before mammals.

- Mammals evolved before birds.

Plan ahead: You will need eight small potted plants for a lab in Chapter 27. If you want to grow your own, start growing radish plants from seed about four weeks before the lab.

Are We Just Relatively Dating?

Chapter 23: Fossil Puzzle Cutouts

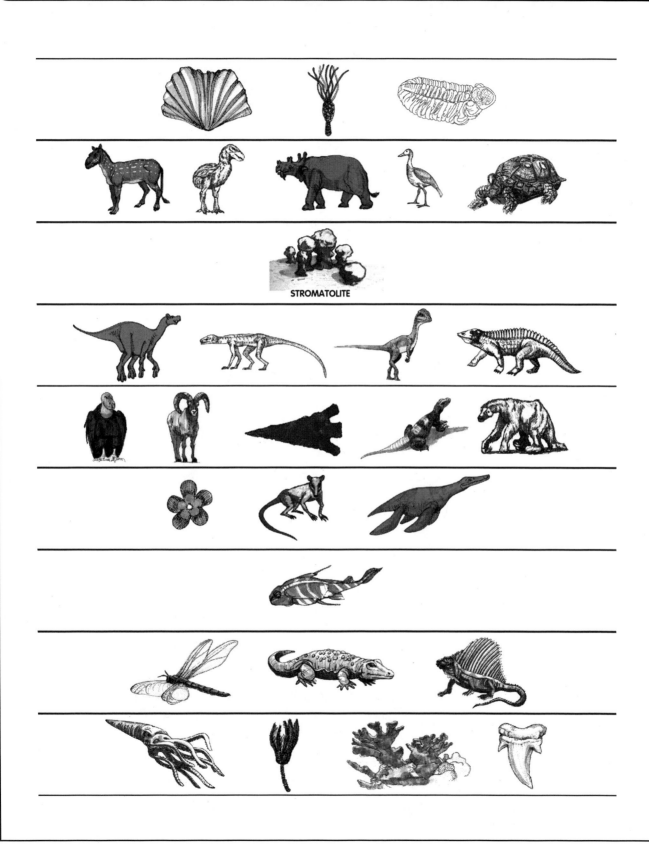

STROMATOLITE

Are We Just Relatively Dating?

Chapter 23: Geological Time Period

CENOZOIC Pleistocene - Holocene - Present
CENOZOIC Paleocene - Pliocene
CRETACEOUS
TRIASSIC - JURASSIC
PERMIAN
MISSISSIPPIAN
DEVONIAN
CAMBRIAN
PRECAMBRIAN

Are We Just Relatively Dating?

Chapter 23: Lab Sheet

Name_____ Date_____

1. What information did you use to place the first puzzle piece?

2. What is the main type of rock where fossils are found in the Grand Canyon?

3. How is this type of rock formed?

4. Locate the Grand Canyon on a map. Why do you think there are both water and land animals fossilized in the Grand Canyon?

5. What principle taught in the chapter does this lab demonstrate?

6. Scientists are not sure exactly how the Grand Canyon formed. They think it formed through erosion, the movement of tectonic plates, and volcanic eruptions. There was a time millions of years ago when there was no canyon; it was just a flat rocky plateau. At that time there must have been fossils in the layers of the rock. What must have happened to the fossils? How does this show a weakness in using the Principle of Superposition alone to date fossils?

Explore

Trees Can Date Too

Chapter 23: Microscope Lab

Scientists look at tree rings to get information about *paleoclimatology*. Paleoclimatology is the study of climates from the past, before accurate records were kept. Scientists use a boring tool to drill into the tree and take a round, cylindrical core sample. They use top-down lighting to do it. With the aid of a computer, they can measure rings to the nearest 0.01 mm.

Trees keep a yearly record. They mark each year of their life with a tree ring in their trunk. Trees record the growing conditions for each year. Thin rings mean the growing conditions were poor with too little water. Thick rings mean the growing conditions were good with lots of water. Scientists use tree rings to study weather patterns, and determine when there have been droughts in the past. A *drought* is a long period without normal rainfall in an area. Droughts affect the growing and living conditions of organisms. Tree rings provide information back 300 years in most areas and thousands of years in some areas.

Today you will look at tree rings with your microscope. Do you think you will be able to tell what years were good for growing and what years were not?

You can tell how old this tree is by counting the rings. The years with thin rings are evidence of years with poor growing conditions. The years with thick rings are evidence of good growing conditions.

Chapter 23: Microscope Lab *continued*

Materials

- Microscope
- Magnifying glass
- Slice of wood that shows tree rings
- Optional: A slide of a slice of wood, purchased if you do not have a source for a slice of wood.
- Sandpaper
- Top lighting
- A piece of paper so pieces of wood do not fall through onto the light in the base

Procedure

1. The slice needs to be smooth. Use sandpaper to smooth it.

2. Look at the slice with a magnifying glass and with a microscope.

3. **For the view with a microscope:** Cover the opening below the stage with paper so no wood chips fall down below. Unless you have a VERY thin slice, do not try to make a slide. Put the slice under the lens and look at it. DO NOT LOOK AT IT WITH POWER HIGHER THAN 40X, unless your slice is very thin. Do not let the lenses rub against the wood. It might scratch them. Draw what you see on your view sheet. Try to analyze what you see.

Trees Can Date Too

Chapter 23: Microscope View Sheet

Name_____ **Date**_____

Specimen _____ Type of mount_____

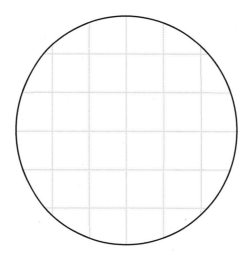

Write an analysis of what you have drawn. Draw lines and indicate years you think were good years with good growing conditions, and those that were poor years with bad growing conditions.

Comments:

Pandia PRESS

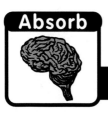

Evolution Act 3: After Darwin
Chapter 23: Famous Science Series

Charles Darwin

Charles Darwin and Alfred Wallace proposed the mechanism of natural selection to explain how organisms evolved. There was one important part missing from the Theory of Evolution as proposed by Darwin. What was it?

When was *The Origin of Species* published?

When was Mendel's paper on genetics published?

Darwin's theory of evolution and the science of genetics have been combined into what is called The Modern Synthesis. The Modern Synthesis gives a more complete understanding of how evolution works. Look up and give a more detailed explanation of The Modern Synthesis.

Many scientists think we are in the midst of a mass extinction right now! What are some of the causes?

How the changes humans are making to the environment are all going to play out, we do not know. As humans continue to change the environment, there will be changes in which traits are superior for the survival of different species. This will result in evolutionary changes in populations. Change in conditions leads to evolution and extinction.

Evolution: When

Chapter 23: Show What You Know

Multiple Choice

1. The study of the layers of sedimentary rock layers is called

   ○ radiometry.
   ○ superposition.
   ○ stratigraphy.
   ○ paleogeography.

2. There are 1,000 carbon-14 atoms in a bone. After two half-lives, 11,136 years, how many carbon-14 atoms will there be?

   ○ 500
   ○ 750
   ○ 250
   ○ 100

3. Tree rings can tell

   ○ about the rainfall in an area.
   ○ the amount of time between good and bad growing years.
   ○ the age of the tree.
   ○ All of the above

4. Carbon-14 is used to date

   ○ igneous rock.
   ○ sedimentary rock.
   ○ metamorphic rock.
   ○ fossilized organisms.

5. Potassium-40 is used to get a relative date for

   ○ dinosaur fossils.
   ○ woolly mammoths.
   ○ metamorphic rocks.
   ○ shells on a beach.

6. Half-life measures

   ○ half the lifetime of carbon atoms.
   ○ 5,568 years.
   ○ the time it takes half the radioactive atoms to turn into stable atoms.
   ○ how long it takes carbon to turn into nitrogen.

7. *(Circle the correct answer)* In a column of rock, a possible progression of fossils is

 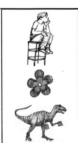 

Chapter 23: Show What You Know *continued*

8. The order of rock layers indicates

   ○ the absolute age of the layers.
   ○ what kind of rock the layer is.
   ○ where the layers were deposited.
   ○ the relative age of the layers.

9. The Principle of Superposition states that

   ○ the oldest layer of rock is at the bottom and the youngest layer is at the top.
   ○ the youngest layer of rock is at the bottom and the oldest layer is at the top.
   ○ all the layers are the same age.
   ○ the position of the layers can be used to give an absolute for fossils.

10. Atoms of an element with different numbers of neutrons in the nucleus are called

    ○ ions.
    ○ molecules.
    ○ isotopes.
    ○ radioactive particles.

11. What isotope of carbon is used for radiometric dating?

    ○ Carbon-12
    ○ Carbon-11
    ○ Carbon-14
    ○ Carbon-13

12. What is the stable isotope carbon-14 turns into?

    ○ Carbon-12
    ○ Phosphorus-40
    ○ Oxygen-16
    ○ Nitrogen-14

13. A layer of igneous rock is running at an angle through layers of sedimentary rock. The igneous rock has an absolute date of 65 million years old. What is the minimum age of the top layer of sedimentary rock through which the igneous rock runs?

    ○ Less than 65 million years old
    ○ More than 65 million years old
    ○ 4.5 billion years old
    ○ A date cannot be found for sedimentary rock because it is made from bits of other rocks.

Chapter 23: Show What You Know continued

Questions

14. What is the name of the principle that explains your answer in question 13?

15. How do changes in environmental conditions affect selective pressures on organisms? Use polar bears and sea ice as an example in your explanation.

Climate change can cause a loss of habitat. The amount of sea ice in the Arctic is decreasing as the global temperature increases. The U.S. Geological Survey Office estimates that there will be two-thirds fewer polar bears in the world by 2050. Polar bears depend on sea ice to find food. If temperatures keep increasing, the amount of polar ice will keep decreasing and polar bears will go extinct or evolve.

Chapter 24: The Biosphere

Biome, Sweet Biome

Chapter 24: Lab

Now, how would you describe where you live? Have you thought about it while you read about the biomes? Today you are going to research and find out the abiotic and biotic factors where you live to determine scientifically which biome is your home.

Materials

- Internet access
- World map

Procedure

1. Fill in the hypothesis on the worksheet. There are five major types of biome. Choose one of these for your hypothesis. If you think you live in the forest biome, choose one of the three types of forest.

2. #1 to #3: Use your computer to fill in the biome worksheet. Search the web for sites that will help determine your longitude and latitude. Try typing these questions into a web search engine: "How far is [my city] from the equator?" or "What is the elevation of [my city]?" A visit to your closest ranger station could also answer many of these questions.

3. #4 to #6: For climate information you can visit the NOAA's National Centers for Environmental Information website, or try searching "climate data for [my city or area]." Write down the maximum average temperature and the month in which it occurs. Then write down the minimum average temperature and the month in which it occurs. Write the annual average for precipitation and snowfall that is at the end of these rows. Then combine them for the total. In chapter 24, snowfall and rainfall were combined for precipitation amounts. This site uses precipitation referring to rainfall. On average, every 10 inches of snow = 1 inch of rain. Divide the snowfall amount by 10 and combine this with rainfall for the total precipitation.

5. Use Google for help with any other questions you have filling in this worksheet.

6. #11: Large bodies of water are abiotic components that can have an effect on a biome. Water absorbs heat, which makes for more moderate temperature fluctuations. Water evaporates into the air, which makes for a moister environment.

7. #12: Choose one answer. Make sure you choose NATIVE vegetation.

Biome, Sweet Biome

Chapter 24: Lab Worksheet

Name_____ **Date**_____

Hypothesis: I think I live in a community that is located in the _____ biome.

Complete the following information for your biome:

1. Longitude _____ Latitude _____

    Distance from the equator in degrees _____

2. Elevation _____ Do you live in the mountains?_____

3. Make a mark on your world map to plot where you live. Use the lines of latitude and longitude to help you plot your location.

4. Average Maximum Temperature: _____ Month it occurred: _____

5. Average Minimum Temperature: _____ Month it occurred: _____

6. Rainfall _____ Snowfall _____ ÷ 10 = _____

    Total precipitation = rainfall + snowfall = _____

7. Is the snowfall close to half or more of the precipitation? _____

8. Are there four distinct seasons or only two? _____

Chapter 24: Lab Worksheet *continued*

9. Do you have a wet and a dry season? _____

10. Are there periodic fires close to where you live? _____

11. Do you live near a large body of water, such as an ocean or a large lake?

12. What is the main type of native vegetation where you live? Circle the one best answer.

   Grassland   Forest   Cactus

   Shrub   Low growing with permafrost on the ground   Algae and seaweed

13. List three animals native to your biome and the adaptations they have for living in it.

Conclusion: Based on your findings, what biome do you live in? Use specific facts to support or correct your hypothesis.

Explore

Diorama

Chapter 24: Activity

Is your favorite biome in your own backyard or far from where you live? Today you will make a diorama of your favorite biome. A diorama is a 3-dimensional display. Have you ever been to a museum and seen a display of a natural history scene? That is a diorama on a large scale. Don't worry; yours won't be that big!

Materials

The materials you use will be different for different biomes. A grassland biome will have lots of grass and only a few trees, but a forest biome will have many trees and just a little grass. The taiga and tundra biomes will both have snow, but the rain forest will not.

- Box – shoebox size or larger
- Nature magazines
- Drawing paper
- Art supplies to draw or decorate the background
- Glue
- Scissors

The following supplies will vary, depending on your biome:

- Animals and plants. You can draw these, make them with Sculpey, or use plastic figures, candy, tissue paper, Legos, etc.
- Pebbles
- For a mountain biome, you can use papier-mâché to make the mountain
- For an aquatic biome, you will want blue paint

Pandia PRESS

Chapter 24: Activity *continued*

Procedure

1. Before you begin, you need to research the biome you wish to create.
   - Make sure you know what plants and animals are in the biome. Include only native plants and animals.
   - What does the ground look like?
   - How much water is available?
   - What is the most common weather?
   - What abiotic features are most important to the biome?

2. Now that you know what your biome should look like, draw a sketch on a piece of paper, showing how you want your diorama to look with the plants, animals, and abiotic features you want in it.

3. Cut the box so there is a bottom and three sides, unless you choose an aquatic biome. If you do, you might want to have a top so you can hang plants and "swimming" animals from the top using string.

4. Decorate the walls of the box with magazine photos or drawings. You can make drawings on paper and glue them to the walls. Or you can paint a base coat on the walls and decorate the walls after the paint dries.

   If you use paint:
   - Work from the back to the front.
   - Paint one wall at a time, with the wall you are painting flat on the ground. Let the paint dry before turning the box.
   - You can paint water, mountains, plants, clouds, a burrow with an animal in it, and many more things. Wait to let the painted surface dry before painting anything on top.
   - Make sure the walls and roof, if you have one, are done before going on to paint the bottom.

5. Think about the weather in your favorite biome when you are decorating the walls. Try to illustrate the weather correctly for the biome.

6. For the ocean biome, show that there is more light at the top of the water than at depth.

7. Decorate the bottom of the box to look like the ground of your favorite biome. If you are using sand or dirt, paint a layer of glue on the bottom of the box, sprinkle the soil on the glue, wait for it to dry, and then lightly shake the loose soil out of the diorama.

8. Decorate your diorama with plants, animals, and physical features like rocks. Place these structures in the diorama before gluing them, so you get them just where you want them. Then, glue them into place. If you have soil on the bottom, you might need a generous amount of glue or to scrape some of the soil away to glue the figures. If you are putting figures in trees or on other structures, attach the figures before gluing them into place in the diorama.

9. Make your figures to scale. For instance, a rabbit will not be bigger than a tree.

10. Write a poem to your favorite biome. The poem follows the tune of the song "Home, Home on the Range." Use the poem instructions and example to help.

A Mad Libs Poem
Chapter 24: Poetry

Fill in the blanks of this poem dedicated to your favorite biome. Make sure you fill the blanks with things that are accurate for the biome. For example, if your favorite biome is the desert, a scorpion would be a good choice for animal, but a polar bear would not. You can write more verses if you like. Read your poem to someone. Copy your poem onto the worksheet provided, or use your own paper to type and illustrate your poem. Tape your biome poetry to the outside of your diorama.

Home, Home in the _____
                                                      your favorite biome

Home, home in the _____
                                          your favorite biome

Where the _____ and the _____ roam
                                plural animal   *plural animal*

Where often is seen a _____
                                                    most common plant in biome

And the sky _____
                              most common weather in biome

Home, home in the _____
                                          your favorite biome

Where animals _____
                                      adaptation used in this biome

Where often is seen a _____
                                                    second most common plant in biome

And the sky _____
                          second most common weather in this biome

Chapter 24: Poetry Worksheet

Name_____ Date_____

Home, Home in the _____

Home, home in the _____,

Where the _____ and the _____ roam.

Where often is seen a _____,

And the sky _____.

Home, home in the _____,

Where animals _____.

Where often is seen a _____,

And the sky _____.

Explore

Soiled

Chapter 24: Microscope Lab

Soil matters. Soil is an important abiotic component of a terrestrial biome. Terrestrial plants need the vitamins and minerals soil contains. Nutrient-rich soil, like that found in grasslands and temperate forests have the nutrients plants need. Soil in some biomes, like the desert and tropical rain forest, is nutrient poor. With all those plants, it seems like rain forest soil would be nutrient-rich, but there is so much rain it leaches the vitamins and minerals out of the soil.

Today you are going to compare different types of soil. Collect as many different types of soil as you can. Soil can be sandy, have clay in it, or be rich with plant material in it. If you gather soil where no plants are growing, predict why.

Materials

- Samples of soil that look visibly different from each other
- Microscope

- Top lighting
- Slides
- Spatula

Procedure

1. Gather soil samples from either outside or from soil mixes for plants. For example, garden stores sell a cactus mix and various mixes that are more and less humus. Try to gather at least one sample from the area where you live.

2. Before making each slide, examine the soil and write your comments and observations about it. Compare the soils in your comments.

3. Use the spatula to put small amounts of soil on each slide.

4. Experiment with the first slide to see if you need top lighting.

5. Look at each sample of soil. View at 40x or 100x and determine which magnification is best. Draw all views from the same magnification. Write a brief description of each sample. DO NOT let the lens of the microscope touch the soil.

Soiled

Chapter 24: Microscope View Sheet

Name_____ **Date**_____

Specimen _____ Type of mount_____

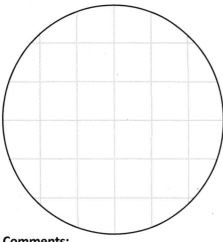

Comments:

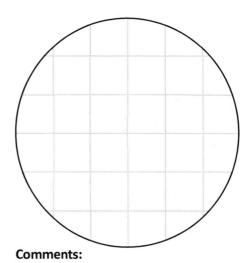

Comments:

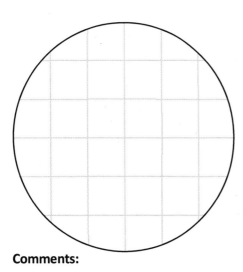

Comments:

Comments:

Pandia PRESS

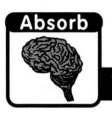

Absorb

Chico Mendes

Chapter 24: Famous Science Series

Working to Save a Biome

Chico Mendes, 1944–1988, was a famous environmental activist and unionist. His story has been told in documentaries, movies, songs, and books.

What biome was he trying to save? Name the area, country, and continent.

Chico Mendez

What was the family business he worked in? How did he go from that to working to save a biome?

In 1987, he flew to Punjab, India, to help stop a project being funded by the Inter-American Development Bank. What was the project he stopped? How do you think this project would have hurt the biome he was trying to save?

Chico Mendes was assassinated. Why?

The Biosphere

Chapter 24: Show What You Know

1. In the blanks, write the abiotic factor most important in creating the biome, WITHOUT PEEKING, if you can. Then circle the adaptations you would expect to find in each of the biomes. Think of this as a logic puzzle. What are the abiotic components important in creating each biome? Which adaptations benefit organisms exposed to those components?

Desert _____

asexual reproduction   deciduous trees   burrowing   hibernation   migration

deep root system   small size   prehensile tail   roots good at absorbing H_2O

specialized kidneys   long arms   leaves to prevent water loss   leaves that shed water

Grassland _____

asexual reproduction   deciduous trees   burrowing   hibernation   migration

deep root system   small size   prehensile tail   roots good at absorbing H_2O

specialized kidneys   long arms   leaves to prevent water loss   leaves that shed water

Tundra _____

asexual reproduction   deciduous trees   burrowing   hibernation   migration

deep root system   small size   prehensile tail   roots good at absorbing H_2O

specialized kidneys   long arms   leaves to prevent water loss   leaves that shed water

Taiga _____

dark green leaves   deciduous trees   long arms   hibernation   prehensile tails

seeds need fire to germinate   subnivean zone   tall trees   migration   burrows

few branches until canopy   fire-resistant bark   leaves that shed water

trees grow close together   trees are thin   trees keep leaves all year

Pandia PRESS

Chapter 24: Show What You Know *continued*

Temperate Forest _____

dark green leaves   deciduous trees   long arms   hibernation   prehensile tails

seeds need fire to germinate   subnivean zone   tall trees   migration   burrows

few branches until canopy   fire-resistant bark   leaves that shed water

trees grow close together   trees are thin   trees keep leaves all year

Rain forest _____

dark green leaves   deciduous trees   long arms   hibernation   prehensile tails

seeds need fire to germinate   subnivean zone   tall trees   migration   burrows

few branches until canopy   fire-resistant bark   leaves that shed water

trees grow close together   trees are thin   trees keep leaves all year

2. Fill in the blank with the best choice from the following words:

ecologists   abiotic   biotic   environment   community   biomes

biosphere   ecosystem   climate   biodiversity   terrestrial   aquatic

_____ are scientists who study the _____. The entire area of Earth where

organisms live is called the _____. It can be divided into areas called _____

that are determined by the _____ of the area. When these areas are on land, they are

called _____. When they are in water, they are called _____.

The biotic and abiotic components in an area form an _____. Climate is controlled by

_____ components, like temperature and precipitation. _____ components

are those that are living or once were. All the organisms that live in an area are a _____.

Areas with a lot of different organisms of many species have a high _____.

Chapter 24: Show What You Know *continued*

3. Multiple Choice

Earth's tilt creates

- ○ temperature extremes at the equator.
- ○ uneven heating of the earth.
- ○ opposite seasons in the Eastern and Western hemispheres.
- ○ ocean waves.

Periodic fires are common in this biome:

- ○ tundra
- ○ aquatic
- ○ tropical forest
- ○ grassland

Because water absorbs sunlight, the highest biodiversity in the ocean is in the ____ zone where ____ takes place.

- ○ sunlight, cellular respiration
- ○ sunlight, photosynthesis
- ○ twilight, cellular respiration
- ○ midnight, photosynthesis

Most autotrophs in the aqueous biome live in the

- ○ river.
- ○ twilight zone.
- ○ sunlight zone.
- ○ lakes.

The difference between freshwater and ocean water is

- ○ freshwater is on land and ocean water surrounds it.
- ○ there are waves in the ocean.
- ○ the amount of dissolved salt.
- ○ latitude.

The location of terrestrial biomes is affected by

- ○ latitude and location on a continent.
- ○ longitude.
- ○ the hemisphere it is in.
- ○ the organisms in it.

An ecosystem can be

- ○ the biosphere.
- ○ a biome.
- ○ a puddle.
- ○ All of the above

The forest biome can be divided into _____ groups. The location of each is determined by its _____ .

- ○ three, amount of snowfall
- ○ two, longitude
- ○ three, latitude
- ○ two, biotic components

The rain forest covers 6% of the earth and _____ of Earth's organisms live there.

- ○ 6%
- ○ 50%
- ○ 15%
- ○ 100%

Why do deciduous trees drop their leaves?

- ○ To maximize opportunities for photosynthesis
- ○ To maximize opportunities for cellular respiration
- ○ To conserve energy and water
- ○ Because it snows

Chapter 25: Predator and Prey

Backyard Food Web

Chapter 25: Lab

Have you ever wondered who was eating whom in your backyard? If you have animals and insects in it, someone is eating somebody; that is a given. In this lab you will learn about the food web that is occurring around you every day. If you don't have a yard, then take a field trip to a place where you can observe nature, like a park, a meadow, or a forest. Only include organisms that occur naturally, however, in your food web. Your dog or cat should not be included. Plants that were planted in your yard might not be included, unless (unfortunately) they have become food for some organism.

Materials

- Colored pencils, optional

- Field guides for the plants and animals in your area (optional)

Procedure

1. Go to your site and study it. Sit quietly to see what is flying or walking around. You do not need to see an animal to include it in your food web, but you do need to know it lives in your environment. For example, if you had a raccoon in your garage a month ago, you know you have raccoons in your ecosystem and you can include raccoons in your food web.

2. Walk around looking at plants and the insects that inhabit them. Look for spiderwebs.

3. Jot down potential candidates for including in your food web. You do not need to document every organism that spends time in your backyard. You can make it as big as you want, but don't let the list get unwieldy. About ten is a good number to document.

Chapter 25: Lab *continued*

4. Start by choosing two or three plants that you know are being eaten. Next, determine what is eating them. It might help to identify the species of organism to assist you in determining what it eats. For example, if you see a squirrel in your food web and you know squirrels eat nuts, but you don't see nuts in your area, maybe they are buried, or maybe the squirrel is just "passing through," or perhaps it's eating something other than nuts. You will have to make these types of determinations based on the evidence you see today or have seen in the past.

5. Record the organisms you identify in your food web on your lab sheet with the following initials (you do NOT need to have all of these in your web).

   - Write P for producers

   - Write C for carnivores

   - Write H for herbivores

   - Write D for a decomposer

   - Write O for omnivores

6. Draw a picture of your food web on the lab sheet. Draw the arrows going away from the organism that is being eaten and going toward the predator. Identify symbiotic relationships, if you observe any:

   - Write CM for organisms practicing commensalism

   - Write M for organisms practicing mutualism

   - Write PS for organisms practicing parasitism

7. OPTIONAL: Color the organisms in your food web.

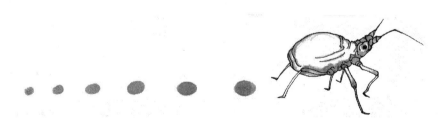

Pandia PRESS

Backyard Food Web

Chapter 25: Lab Sheet

Name_____ **Date**_____

Organisms in my food web:

My food web

Key: P = Producer   C = Carnivore   H = Herbivore   D = Decomposer   O = Omnivore

CM = Commensalism   M = Mutualism   PS = Parasitism

 **Explore**

Plant Predation

Chapter 25: Microscope Lab

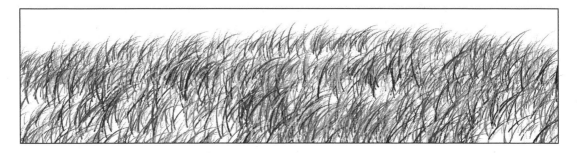

The predation of plants, such as grass, is called *grazing*. Like other forms of predation, most grazing is helpful in maintaining diversity in the community. But it can be harmful to the organism that is the prey. Let's look at a plant before and after it has been preyed upon by you. Grazers bite on grass and tear it. You will use pliers to simulate the teeth of a grazer.

Materials

- Microscope
- Slides
- Water

- Slide cover
- One piece of freshly picked green grass
- Pliers

Procedure

1. Make a wet mount slide with the blade of grass at a 90-degree angle to the slide. It will hang over the sides. View the piece of grass at a place where it was not picked. This is the Before view. Find the magnification you prefer and draw this view. Remove the slide and take off the slide cover.

2. At about the point you viewed with the microscope, take the pliers, squeeze them on the blade of grass, and tear it in two pieces.

3. Make a wet mount slide so you can view the site of the tear. This is the After view. Draw this view at the same magnification you drew the Before view. Remove the slide and take off the slide cover.

 Pandia PRESS

Plant Predation

Chapter 25: Microscope Lab Sheet

Name_____ **Date**_____

Specimen _____ Type of mount_____

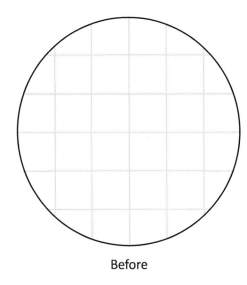

Before

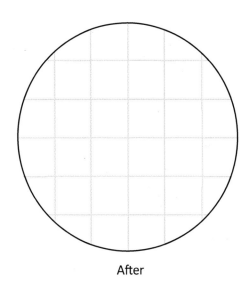

After

Describe the differences between the two blades of grass.

Why do you think grazing can be harmful to the organism to which it happens?

Sometimes when grass is cut, the tip turns brown. Use the differences you observed between the two blades to explain why this might occur.

Absorb   Jane Goodall

Chapter 25: Famous Science Series

A good title for Jane Goodall could be "Famous Observer of Biotic Interactions." What organisms did Ms. Goodall observe, and why is this a good title?

Jane Goodall

What are the two main threats that chimps face today?

It has been said that Jane Goodall, through her work, has changed the way we think about all animals. She once said, "Only if you understand, will you care. Animals have feelings too." What did she mean by this?

Pandia PRESS

Chapter 25: Famous Science Series *continued*

Many of the volunteer and outreach programs started by Goodall reach out to kids your age. How can people your age help save chimps?

A famous anthropologist gave Jane Goodall her start. What is his name and why is he famous?

The Gombe National Park is the location of the chimps Goodall studied. What country is Gombe National Park in?

What lake does the park border? Find the lake and Olduvai Gorge, where Leakey made his discoveries, on a map of Africa.

What famous explorer discovered the lake's only outlet?

At Gombe National Park, humans need to keep 10 meters away from the chimps. Why?

Predator and Prey
Chapter 25: Show What You Know

Questions

1. If you have a pet, look at the list of ingredients on your pet's food. Write down the ingredients here and answer this question: Is your pet an omnivore, herbivore, or carnivore? If you do not have a pet, look at a bag of pet food the next time you are at the grocery store.

2. Draw a food web using the organisms below. You do not have to use all of them. Draw the arrow going away from an organism to what it might be eaten by. Mark organisms that make their own food with a P, for producer.

Chapter 25: Show What You Know *continued*

3. List three adaptations predators have for catching prey.

4. List three adaptations prey animals have for avoiding being caught. Explain the benefits of each adaptation.

5. Multiple Choice

Cleaning fish will go into the mouth of a barracuda and clean its teeth, eating any parasites they find. This is an example of

- ○ predation.
- ○ symbiosis.
- ○ mutualism.
- ○ All of the above

Going from producer to herbivore to carnivore:

- ○ There is more energy
- ○ There is less energy
- ○ There is the same amount of energy

When two species have a similar niche, they use _____ to reduce competition.

- ○ resource partitioning
- ○ a predator/prey relationship
- ○ intraspecific interactions
- ○ mimicry

Coral snakes have yellow, red, and black stripes. This is an example of

- ○ bioluminescence.
- ○ camouflage.
- ○ disruptive coloration.
- ○ aposematic coloration.

Commensalism is a symbiotic relationship where

- ○ both species benefit.
- ○ neither species benefits.
- ○ one species benefits and the other is harmed.
- ○ one species benefits and it doesn't affect the other species.

The most intense competition is

- ○ intraspecific competition.
- ○ interspecific competition.
- ○ between predators and their prey.
- ○ between species that mimic each other.

Chapter 25: Show What You Know *continued*

Plants defend themselves against predators using

- ○ chemicals.
- ○ hard shells to protect seeds.
- ○ thorns.
- ○ All of the above

An example of intraspecific competition is

- ○ a dog chasing a cat.
- ○ a dog marking its territory.
- ○ a dog barking at a strange person.
- ○ a dog eating grass.

The predator/prey relationship is beneficial to a community because

- ○ they are cool to look at.
- ○ they limit the damage done to plants.
- ○ they increase the diversity in the community.
- ○ they decrease the diversity in the community.

A population's niche is its

- ○ feeding strategy.
- ○ job in the community.
- ○ adaptation.
- ○ type of symbiosis.

Questions

6. Plants are called producers because they produce their own food. What is the name of the process they use to do this? In what organelle does this process occur? Write the chemical reaction and state the name of the food made in this process. (10 points if you do not have to peek, 5 if you do)

7. All organisms need energy. What is the name of the process used to make energy? In what organelle does this process occur? Write the chemical reaction for this process.
   (10 points if you do not have to peek, 5 if you do)

Chapter 26: Cycles

Cycling Resources
Chapter 26: Lab

Today you will build an ecosystem. It is a closed ecosystem, called a terrarium. Your terrarium will have all the abiotic and biotic components it needs to sustain life. The organisms in your terrarium will be plants that you can see and microorganisms, like fungi and bacteria, that you cannot see without a microscope.

Your terrarium will be closed so that carbon, nitrogen, phosphorus, and water cycle through it. Water is made from hydrogen and oxygen, and they will cycle through the ecosystem you created too. You will be able to observe some parts of the water cycle directly. You will add water to the terrarium when you are making it. When the enclosed terrarium is exposed to warmth and sunlight, the water will evaporate, condense,

precipitate, and percolate. That means it will rain in your terrarium! In fact, all of the four biogeochemical cycles discussed in this chapter will happen in your terrarium. Osmosis, transpiration, and photosynthesis will also happen in your terrarium; they are not as easy to observe, though. These cycles cannot be observed as directly as the rain-making part of the water cycle. But you will know that they are occurring because the plants in your terrarium will grow. Can you think what it means if plants grow? It means the plants are getting the molecules they need to make cells. Five of the elements and one of the molecules plants need to make cells are carbon, nitrogen, phosphorus, oxygen, hydrogen, and water. The terrarium you create has all of these. Do you remember what the other element is that organisms need to make cells? If you answered calcium, you remembered correctly, and your terrarium must have that, too, or the plants won't grow.

Suggestions and Warnings:
★ Do not put your terrarium in full sun.
★ Choose small, slow-growing plants.
★ Make sure all abiotic components have been cleaned in hot water (or you may get insects or disease).
  • Use potting soil that has been pasteurized.
  • Carefully inspect plants for insects.
  • Do not add fertilizer.
  • Carefully watch your terrarium for the first week. It is critical that you get the amount of sunlight and water just right. You know you have gotten it right if it rains in your terrarium without getting soggy.
  • You might need to prune the plants in your terrarium so none take it over.

Chapter 26: Lab *continued*

Materials

The amounts of the materials you get will depend on the size of your container.

- Water
- Container with an <u>airtight</u> lid. Choose this first so you know how much of the other material to get. Your container can be as big as an aquarium or as small as a mason jar. Try to find a container with a large enough opening that your hand fits inside of it.

- Pea gravel or pebbles that have been cleaned in hot or boiling water, enough for a 2-inch layer
- Activated charcoal like that used in aquarium filters, enough for a ¼-inch layer
- Sphagnum moss, also called Spanish moss, enough for a 1-inch layer

- Soil (do not use sand), enough for a 3- to 4-inch layer
- Plants. Choose plants that have the same moisture and light needs. There is a list of good plants for terrariums on the next page. Do not crowd plants.

Procedure

1. Make sure everything you use is clean!

2. You will build a four-layer foundation for your plants.
   - Spread a 2-inch layer of pea gravel or pebbles—this is for drainage
   - Spread a ¼-inch layer of activated charcoal on the gravel—this is to clean the air of fumes when organic material decomposes
   - Spread a 1-inch layer of sphagnum moss—this keeps the soil from falling through to the bottom
   - Spread a 3- to 4-inch layer of soil

3. Plant all the plants except mossy plants, if using, spaced carefully around your aquarium. When all the plants are planted, sprinkle a layer of soil around them and tamp it down.

4. Plant any moss plants now, if you are using.

5. Slowly and carefully, pour cool water into the terrarium until you see it *begin* to pool in the rock layer.

6. Put the lid on tightly.

7. Put the terrarium in a location that gets moderate amounts of indirect sunlight.

8. Check your terrarium the next day. Look for a little to a medium amount (NOT A LOT) of condensation. This will show up as water beading up on the inside of the container.

9. Draw a picture of your terrarium in the space below. On your drawing, map the water cycle that is happening in your terrarium ecosystem.

Chapter 26: Lab *continued*

10. Terrarium maintenance:

- You may need to water your terrarium every few weeks. A spritz bottle is good for this.
- If you do add water, add a small amount.
- Expect droplets to form on the sides of the jar, but if it looks very foggy, take the lid off for a couple of hours and let it dry. You want it to rain but you do not want to saturate.
- Do not have soggy soil!
- Overwatering can lead to root rot, or the growth of fungus, or mold.
- If a plant looks diseased or funky, remove it right away.
- Prune plants that get too big.

Plants for Terrariums Choose SMALL plants (it's acceptable to use all moss, if you like).

Some good plants to use: miniature spider plant, earth star, Irish moss, miniature fern, nerve plant, button fern, creeping fig, Swedish ivy, bird's nest sansevieria, maidenhead spleenwort, miniature peperomia, miniature African violet, creeping fig, baby tears

My Terrarium Ecosystem

Explore

Let's Have a Symbiotic Relationship
Chapter 26: Microscope Lab

Rhizobacteria are a type of bacteria that take nitrogen out of the atmosphere and turn it into nitrogen that plants can absorb through their roots. The process of changing atmospheric nitrogen into a different chemical is called nitrogen fixing. Types of plants, called legumes, have a symbiotic relationship with rhizobacteria. Plants that are legumes include beans, lentils, peas, and peanuts. The rhizobacteria provide nitrogen for the plant in exchange for a safe place around the roots of the plant to live and reproduce. What type of symbiotic relationship benefits both species?

Nurseries and hardware stores sell inoculant for legumes. Legume seeds are coated with legume inoculant before planting them. Inoculant contains rhizobacteria. Today you will look at rhizobacteria with your microscope. Do you think you will see them fixing nitrogen?

Materials

- Microscope
- Slides
- Water

- Slide cover
- Legume inoculant (rhizobacteria)

Procedure

Make a wet mount slide of the legume inoculant. Wait 15 minutes then look at it. Draw a view of the rhizobacteria at your favorite magnification. Write your comments about what you saw.

Let's Have a Symbiotic Relationship

Chapter 26: Microscope Lab Sheet

Name_____ **Date**_____

Specimen _____ Type of mount_____

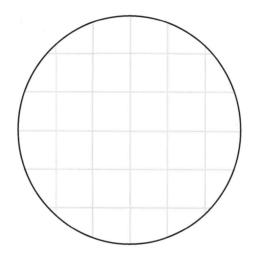

Comments:

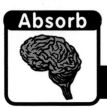

Charles David Keeling
Chapter 26: Famous Science Series

Famous Carbon Cycle Pioneer

Charles David Keeling (April 20, 1928 – June 20, 2005) was a passionate advocate for the environment.

What did Keeling begin measuring in 1958?

Where did he make these measurements?

In 1961, data collected by Keeling showed that the amount of carbon dioxide in the air was increasing. When he made a graph from the data it showed a curve that has been named the Keeling Curve. Draw a quick sketch of the Keeling Curve below. Remember to label the axes.

Chapter 26: Famous Science Series *continued*

Based on the Keeling Curve, in the near future, do you think the amount of carbon dioxide molecules in the air will increase or decrease?

Did Keeling show that the increase in carbon dioxide was from artificial sources or from natural sources?

What award did Keeling win in 2002?

Cycles

Chapter 26: Show What You Know

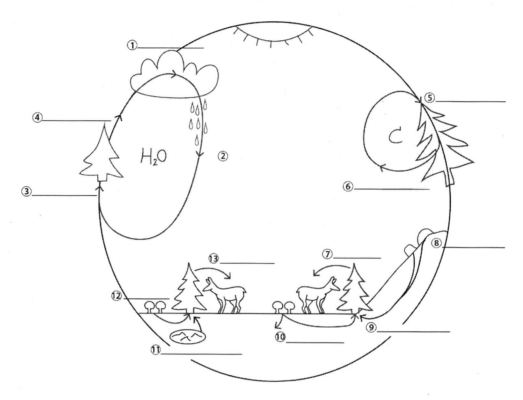

Identify the biogeochemical cycles pictured above. In the left-hand column, name the four cycles.
In the right-hand column, name the processes occurring within each cycle.

_____ {
1. _____
2. _____
3. _____
4. _____

_____ {
5. _____
6. _____

_____ {
7. _____
8. _____
9. _____
10. _____

_____ {
11. _____
12. _____
13. _____

Chapter 27: Threats

How Acids Affect Life

Chapter 27: Lab

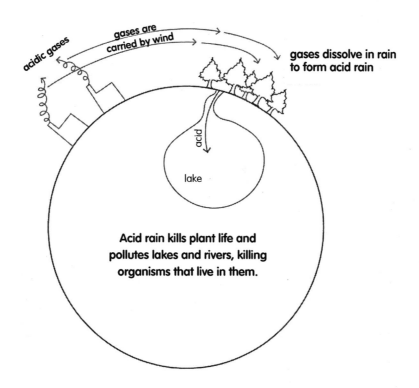

Acid rain kills plant life and pollutes lakes and rivers, killing organisms that live in them.

In addition to carbon dioxide, the burning of fossil fuel also releases nitrogen oxides and sulfur dioxide into the atmosphere. These gases react with water vapor in the air and form nitric and sulfuric acids. These acids fall to the ground in acidic precipitation called acid rain. Acid rain is an environmental threat wherever it falls. It has caused the most damage to the forest biome and the aqueous biome by making soil and water acidic. This affects the chemical balance of the soil and can kill organisms exposed to the acid.

For this experiment, you will water four separate plant groups with four different watering solutions, ranging in concentration from water only to vinegar only. Vinegar is an acid. The higher the concentration of vinegar in solution, the more acidic the solution is.

This lab takes one to three weeks from start to finish. The length of time the experiment takes to conduct depends on variables, such as the size of the plants and how sensitive the plant species you choose is to acid.

Chapter 27: Lab *continued*

Materials

- 8 small potted plants of the same type, e.g., 8 radish plants. Pots should have holes for drainage.
- 8 dishes, 1 for under each pot

- 4 jars with lids that each hold 2 cups of liquid
- Tablespoon
- Marking pen
- Potting soil

- 3½ cups white vinegar
- 12 marking tags (or labels)
- 4¾ cups distilled water
- Measuring cup
- Camera (optional)

Procedure

1. Label the potted plants and jars. Write the following on the 12 labels:

- On three labels write "H_2O" indicating only water. Put two of the labels on two potted plants, and one on a jar.

- On three labels write "4 H_2O to 1 V." This means four parts water to one part vinegar. Put two of the labels on two unlabeled potted plants, and one on an unlabeled jar.

- On three labels write "1 H_2O to 1 V." This means one part water to one part vinegar. Put two of the labels on two unlabeled potted plants, and one on an unlabeled jar.

- On three labels write "V," indicating only vinegar. Put two of the labels on the last two unlabeled potted plants, and one on the last unlabeled jar.

2. Make four solutions, one in each jar.

- In the jar labeled H_2O, pour in 2 cups of water.

- In the jar labeled 4 H_2O to 1 V, measure 12 ounces (this is 1½ cups) of water and add 3 ounces of vinegar. This is a total of 15 ounces. There are 16 ounces in 2 cups of liquid. For the last ounce you will use the relationship: 6 teaspoons = 1 ounce. Measure and add 4 teaspoons of water and 1 teaspoon of vinegar into the jar. The volume in this jar is 1 teaspoon less than 2 cups, but it will be okay.

- In the jar labeled 1 H_2O to 1 V, measure 1 cup of water into jar and add 1 cup of vinegar.

- In the jar labeled V, measure 2 cups of vinegar.

3. Matching the label on the pot with the label on the jar, water each potted plant every day with ⅛ of a cup of liquid. Start watering today on day one. Water for 14 to 21 days, or until you have conclusive results. Do not under-water.

4. After each watering, put the lids on the jars. Do not mix lids to prevent cross-contamination.

5. After watering on day one, write your hypothesis on the chart for each potted plant. Predict how you think the plant will respond to the solution it is being watered with.

6. Make notes about your observations every day on your lab sheet.

7. Optional: Take pictures every day of the experiment to include with your lab.

8. After 21 days, or when you think you have enough data, write a conclusion. To which solution did the plants respond the best? Which resulted in the worst response? Based on your results, how does acid affect plant life?

9. Write a formal report for this lab on the Lab Report sheet.

How Acids Affect Life

Chapter 27: Lab Sheet

Name_____ Date_____

Watered With	Hypothesis	Days	Week One	Week Two	Week Three
Water only		Day 1			
		Day 2			
		Day 3			
		Day 4			
		Day 5			
		Day 6			
		Day 7			

Watered With	Hypothesis	Days	Week One	Week Two	Week Three
Four parts water to one part vinegar		Day 1			
		Day 2			
		Day 3			
		Day 4			
		Day 5			
		Day 6			
		Day 7			

Chapter 27: Lab Sheet *continued*

Watered With	Hypothesis	Days	Week One	Week Two	Week Three
Half water, half vinegar		Day 1			
		Day 2			
		Day 3			
		Day 4			
		Day 5			
		Day 6			
		Day 7			

Watered With	Hypothesis	Days	Week One	Week Two	Week Three
Vinegar only		Day 1			
		Day 2			
		Day 3			
		Day 4			
		Day 5			
		Day 6			
		Day 7			

Conclusion:

Pandia PRESS

How Acids Affect Life

Chapter 27: Lab Report

Name_____ Date_____

Hypothesis

Observations

Results and Calculations

Conclusions

 **Acids, Up Close and Personal**
Chapter 27: Microscope Lab

This lab is started the day before observations can be made.

Acids are damaging to life at the cellular level. Do you think you will be able to see the damage with your microscope?

Materials

- Microscope
- 2 small, nonmetallic bowls
- ½ cup distilled water
- ½ cup white vinegar

- 2 leaves from the same source
- Measuring cup
- X-Acto knife
- Slides

- Slide covers
- Water to make a wet mount slide
- Syringe

Procedure

1. Measure ½ cup of vinegar into one bowl.

2. Measure ½ cup water into one bowl.

3. Put one leaf in each bowl.

4. Let them sit overnight.

5. The next day, cut a small section from each leaf. Make two wet mount slides, one for each leaf section.

6. Draw a picture of each at 100x magnification. Write your observations.

Acids, Up Close and Personal

Chapter 27: Microscope Lab Sheet

Name_____ **Date**_____

Specimen _____ Type of mount_____

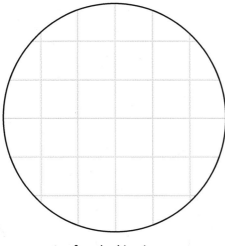

Leaf soaked in vinegar

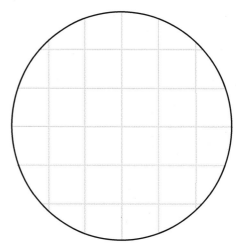

Leaf soaked in water

Observations:

Is acid good for leaves?

You already knew water was not harmful to leaves. You included it as a control for this experiment. What does that mean and why is it important to have controls in experiments?

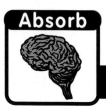

Rachel Carson
Chapter 27: Famous Science Series

Fighting to Save Planet Earth

Rachel Carson (1907 – 1964) was a writer, marine biologist, and ecologist.

Carson is credited as one of the people who sparked the environmental movement in the United States.

Carson wrote a famous book entitled *Silent Spring*. The purpose of the book was to enlighten people about what environmental problem?

Carson grew up near a city and two rivers that, at that time, were very polluted. What were the names of the city and rivers, and what was the industry causing the pollution?

Rachel Carson

Carson wrote an article and submitted it to a magazine when she was 10 years old. The magazine published it. What was the name of the magazine?

Carson worked for what agency in the United States government?

What was the name of the pesticide Carson writes about in *Silent Spring*?

What U.S. president read Carson's book and created a panel to investigate her claims?

In 1970, the U.S. government created the EPA. What does this acronym stand for? In 1972, the EPA banned the use of what pesticide in the United States?

Pandia PRESS

Threats

Chapter 27: Show What You Know

Multiple Choice

1. Acid rain

- ○ is a fertilizer that promotes growth in forests.
- ○ helps crops grow by killing insects.
- ○ helps crops grow by killing competing plants.
- ○ makes soil and water acidic which kills organisms in the soil and water.

2. A habitat is

- ○ where a house mouse lives.
- ○ part of a biome.
- ○ the kind of environment an organism can live in.
- ○ fixed for species.

3. Climate change

- ○ is the same everywhere.
- ○ is worse at the equator.
- ○ affects each biome differently.
- ○ has increased the rainfall in the desert biome.

4. Pesticides, herbicides, and fertilizers can harm the environment because

- ○ they change the chemistry of the environment.
- ○ they make things grow in water.
- ○ they kill weeds but not crops.
- ○ All of the above

5. When the water around coral reefs warms,

- ○ the coral thrives.
- ○ desertification can occur.
- ○ the photosynthetic algae leaves the coral polyp, and the polyp dies.
- ○ coral polyps are more susceptible to eutrophication.

6. The number of animals poached from the rainforest every year is

- ○ about 1,000,000.
- ○ about 10,000.
- ○ about 100,000.
- ○ over 10,000,000.

7. Salmon ladders are one example of

- ○ loss of habitat.
- ○ a device people have created to return habitat back to salmon.
- ○ part of a water park.
- ○ a type of dam.

8. Farming, deforestation, and dams all result in

- ○ loss of habitat.
- ○ desertification.
- ○ eutrophication.
- ○ extirpation.

Chapter 27: Show What You Know *continued*

9. The term *carbon footprint* refers to

○ a trace fossil.

○ the things each person does that produce carbon dioxide.

○ the fact that humans are carbon-based life forms.

○ None of the above

10. Littering and the dumping of trash in the ocean has led to

○ climate change.

○ eutrophication.

○ a floating garbage patch that is the size of Texas off the coast of Hawaii.

○ All of the above

Matching. Match the word with its definition.

Eutrophication ○   ○ When non-desert areas turn into desert

Extirpation ○   ○ Illegal hunting or catching of organisms

Poaching ○   ○ When so many fish are caught that the population cannot sustain itself

Overgrazing ○   ○ Fertilizer makes the plants in aqueous ecosystems grow better

Desertification ○   ○ When livestock eat so much grass that the grass cannot sustain itself

Overfishing ○   ○ The local extinction of a population

Deforestation ○   ○ When forests are cut down

Pandia PRESS

Chapter 27: Show What You Know *continued*

Short Answer

List 5 things you could do to reduce your carbon footprint.

1.

2.

3.

4.

5.

Think of an environmental problem that is affecting the area where you live. Make a plan for what you can do to help.

Environmental threat in my area:

A plan to fix it:

Chapter 28: Taxonomy

Dichotomous Key Mystery
Chapter 28: Lab

Sherlock Watson, the famous collector of qwitekutesnutes, has died. His collection has sixteen qwitekutesnutes. No two are the same. In his will, he has left each of his qwitekutesnutes to a different zoo. You are given the job of determining where each qwitekutesnute will be sent. When you get there, you are surprised to find the qwitekutesnutes together playing. They are not in individual cages. Oh, no! Which qwitekutesnute is supposed to go to which zoo? Then you notice a note on the table.

If you are reading this, you must be the person responsible for sending my beloved qwitekutesnutes to their new homes! I have decided to leave one last mystery. You will have to determine where each qwitekutesnute is to be sent. To help you solve this mystery I have left you a dichotomous key.

Happy detecting, Sherlock Watson

In biology, **dichotomous keys** are used to identify unknown organisms and to classify new species of organisms. A dichotomous key uses differences to identify something in a systematic way. *Dichotomous* means "divided in two parts." At each step in the key, there are two choices to make, for example, white fur and black fur. When enough choices are completed, an identification is made.

After you use Sherlock Watson's dichotomous key to send all the qwitekutesnutes to the correct zoos, you will make your own dichotomous key. Are you as good at solving and writing mysteries as Sherlock Watson? Are your friends and family as good at solving them as you are?

Materials

- Scissors
- Glue or tape
- Friends to try solving your dichotomous key

- Magnifying glass
- 4 envelopes

- 12 leaves, 2 each from 6 different species of plant. (If the leaves are from similar species of plants, such as all conifer trees, your dichotomous key will be more challenging to solve.)

Chapter 28: Lab *continued*

Procedure

1. On the Qwitekutesnute Sheet, cut the 16 qwitekutesnutes into 16 cards.

2. Use the dichotomous key below to work out which qwitekutesnute goes to each zoo. At each level starting from the top, separate the qwitekutesnutes into groups matching the description for that trait. For example, separate them into two piles—one pile for the qwitekutesnutes with stars and one pile for the qwitekutesnutes with stripes. Then take one of these piles and separate it into two piles—one pile for those with a Mohawk and one pile for those with ear tufts. Do this for each pile until you get to a point where there is only one qwitekutesnute. The letter in parenthesis indicates a qwitekutesnute's zoo destination.

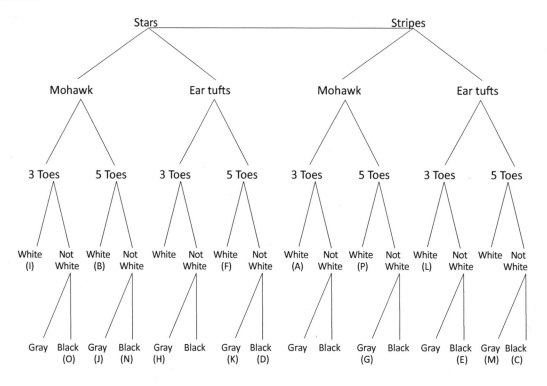

3. Glue or tape each qwitekutesnute under the correct zoo name on the Zoo Destinations lab sheet.

4. Examine the leaves with your eye and with a magnifying glass. Notice at least four differences between the leaves. There will probably be more differences than four. Include at least one difference that you need a magnifying glass to see.

5. On the My Leaf Dichotomous Key lab sheet, make a dichotomous key for the leaves by writing two choices to be made at several junctures. Start at the top and write a difference that applies to all the leaves in the blanks provided (e.g., smooth-texture or rough texture). Divide the leaves into these two groups, then starting with one group (i.e., smooth texture leaves) make more junctures that illustrate the differences until you run out of leaves. Now do the same with the other group (e.g., rough texture). You must make sure that no more than one leaf fits in each place on your dichotomous key. (Some places might have no leaves, and some paths might end before you get the bottom.)

6. Label the envelopes A, B, C, D, E, and F. Put one leaf in each envelope. Have friends and family choose an envelope and see if they can solve your dichotomous key for that leaf.

Chapter 28: Qwitekutesnute Sheet

Dichotomous Key Mystery

Chapter 28: Lab Sheet – Zoo Destinations

Name_____ **Date**_____

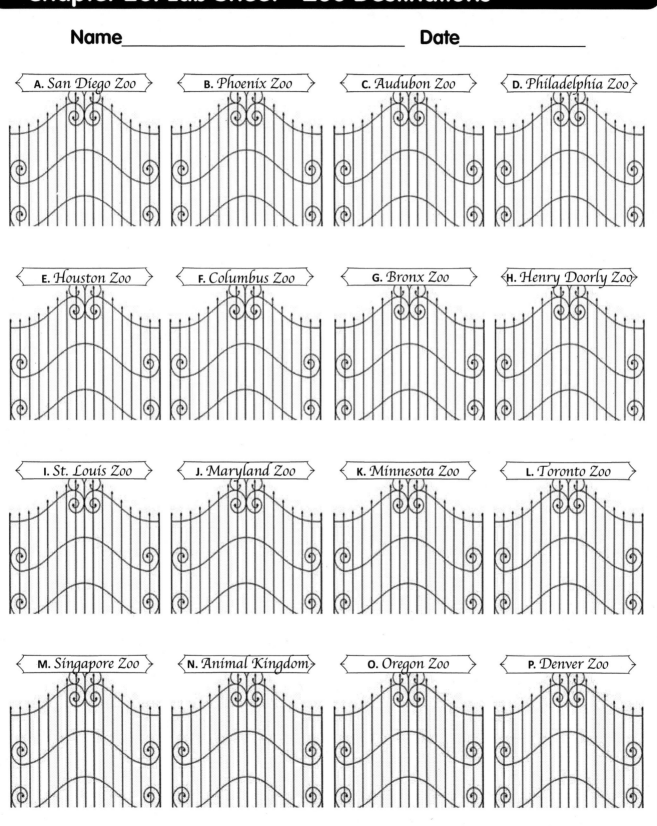

A. San Diego Zoo   **B.** Phoenix Zoo   **C.** Audubon Zoo   **D.** Philadelphia Zoo

E. Houston Zoo   **F.** Columbus Zoo   **G.** Bronx Zoo   **H.** Henry Doorly Zoo

I. St. Louis Zoo   **J.** Maryland Zoo   **K.** Minnesota Zoo   **L.** Toronto Zoo

M. Singapore Zoo   **N.** Animal Kingdom   **O.** Oregon Zoo   **P.** Denver Zoo

Dichotomous Key Mystery

Chapter 28: Lab Sheet – My Leaf Dichotomous Key

Name_____ **Date**_____

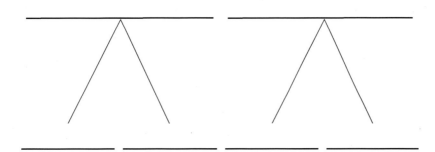

Absorb

Carolus Linnaeus
Chapter 28: Famous Science Series

The Famous Taxonomist

When and where was Carolus Linnaeus born?

Linnaeus invented the system for naming organisms because there was no standard. One organism might have several different scientific names. Many of the names were long with many words. He streamlined that to two words: the binomial system. Why was this needed?

Carolus Linnaeus

What name did he give humans? What does that name mean?

Linnaeus proposed a system of classification with three kingdoms. What were they?

Why don't you think he proposed a kingdom for bacteria?

Pandia PRESS

Chapter 28: Famous Science Series *continued*

Who proposed the five-system kingdom? When was it proposed? What were the names of the five kingdoms?

One of the kingdoms was split into two kingdoms. Which kingdom was it and why was it changed?

Discussion question: Explain what is meant by this statement:

The system of classification proposed by Linnaeus changed over time. It is a good example of how a scientific theory can grow and change as new information becomes available.

Species Spotlight
Chapter 28: Research

Earth is full of different species. You will choose one of them, research what is known of its biology, and write about what you learned.

Materials

- Computer and printer, or paper and pen
- Access to the Internet and/or a library

- Essay Worksheet (Appendix workbook)
- Attribution of Sources (Appendix main text)

Procedure

1. Choose a species.

   - It can be any species of organism alive today. Choose a species, not a genus or a family. Keep it specific. Choose one that fascinates you.

   - Do a little research after you choose a species, to make sure there is enough information about that animal for a full-fledged report. If you are having trouble finding information, be willing to change the species to one with more information available.

   - Use at least five sources for your research report (at least two of these should be print sources). Read Attribution of Sources and type or write a bibliography of the sources you used.

2. Complete the Research Report form for the species you chose. When illustrating, you can draw pictures or paste a photograph. For evolutionary history, describe when and where the organism evolved and its ancestral species. For the current status of the species, explain if it is thriving or endangered. If endangered or threatened, explain why and describe what can be done, or is being done, to save it.

Pandia PRESS

Chapter 28: Research *continued*

3. Write an essay about the species. Use the Essay Worksheet as a guide when writing your essay. Choose from these two essay assignments:

   1. Write a three- to five-paragraph essay detailing the facts you learned about your organism.

   2. Write a three- to five-paragraph fictional story about your organism. Make sure you have at least five facts about your organism woven into the story.

4. Start writing. If you feel really creative, you could even write a poem about the species you are researching. Unless otherwise instructed by your teacher, you should continue with this course while you are researching and writing your report. Ask your teacher to review and comment on a rough draft before composing your final report.

5. Create a cover sheet for your essay report and attach your bibliography.

Species Spotlight

Chapter 28: Research Report

Name_____ Date_____

Title

Scientific Name	**Illustration of the Species**
Classification of the Organism:	
Domain:	
Kingdom:	
Phylum:	
Class:	
Order:	
Family:	
Genus:	
Species:	

Location. Color and label where the species lives.

Biome. Describe in words and/or pictures the biome(s) the species lives in.

Chapter 28: Species Report *continued*

Life History

Life span:

Method of reproduction:

Number of offspring:

Description

Size:

Adaptations:

Niche:

Nutrients

Its primary food and how the species feeds:

Place the species in a food web (describe or illustrate):

Evolutionary History

Current Status of the Species

Taxonomy

Chapter 28: Show What You Know

Classification. List the eight levels of classification in order. Start with the level that has the most species in it and end with the level that has the fewest species in it.

1.   5.

2.   6.

3.   7.

4.   8.

Multiple Choice

1. The three domains are

   ○ Archaea, Bacteria, Eukarya.
   ○ Kingdom, Genus, Species.
   ○ Plantae, Animalia, Fungi.
   ○ Family, Genus, Species.

2. Shared traits are

   ○ traits that are shared by organisms of the same species.
   ○ symbiotic.
   ○ traits that were inherited from a common ancestor.
   ○ when species have the same feeding strategies.

3. The branch of biology that classifies and names organisms is

   ○ binomial classification.
   ○ domain system.
   ○ dichotomous nomenclature.
   ○ taxonomy.

4. The naming system Linnaeus developed for naming species is called

   ○ binomial nomenclature.
   ○ domain system.
   ○ dichotomous classification.
   ○ taxonomy.

5. Modern taxonomy uses

   ○ genetic evidence and computers.
   ○ anatomy.
   ○ methods of feeding.
   ○ All of the above

6. Which is the correct way to write the scientific name for a koala bear?

   ○ phascolararctos cinereus
   ○ <u>phascolararctos</u> <u>cinereus</u>
   ○ Phascolararctos cinereus
   ○ <u>Phascolararctos</u> <u>cinereus</u>

Chapter 28: Show What You Know *continued*

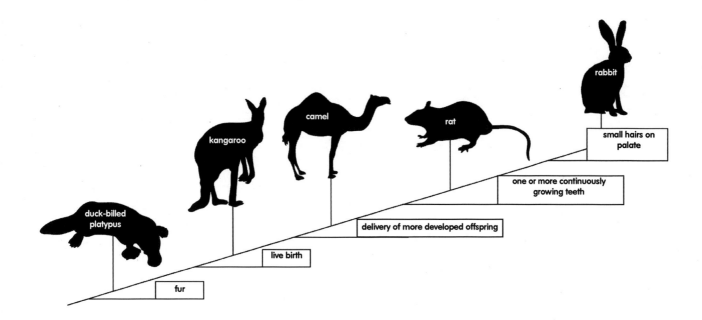

Diagram. Use the diagram above to answer questions 7–10.

7. The name of the diagram is a

    ○ dichotomous tree.

    ○ dichotomous key.

    ○ cladogram.

    ○ taxonomy.

8. The rabbit is most closely related to the

    ○ rat.

    ○ camel.

    ○ kangaroo.

    ○ duck-billed platypus.

9. Which animal has all the shared traits listed?

    ○ Rabbit

    ○ Rat

    ○ Camel

    ○ Kangaroo

    ○ Duck-billed platypus

10. Which shared ancestral trait is shared by all animals?

    ○ Fur

    ○ Continuously growing teeth

    ○ Live birth

    ○ Small hairs on the palate

Chapter 28: Show What You Know *continued*

Cladograms. Fitting the information into a cladogram is a logic puzzle. There are three cladograms for you to work—simple, basic, and advanced. Fit the organisms and traits into the spaces on the cladograms. If you aren't sure if an organism has a trait, look it up on the Internet or in an animal encyclopedia. Begin by trying to find the traits all the organisms share. For example, start by asking if all the organisms are heterotrophs. If yes, it goes on the bottom horizontal line. If no, work your way down the list until you find the trait all the organisms share. Next ask yourself which organism has that trait but does not have the other trait. Write that organism's name at the top of the lowest vertical line. Repeat this over and over until the cladogram is complete.

Simple Cladogram

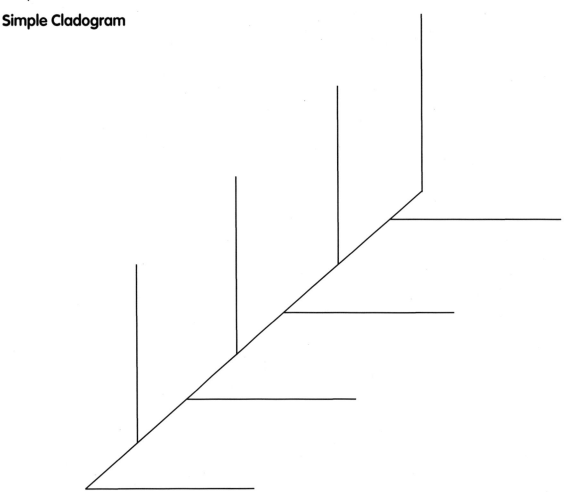

Organisms
1. Owl
2. Palm tree
3. Shark
4. Penguin

Shared Traits
1. Heterotroph
2. Feathers
3. Ability to fly
4. Multicellular

Pandia PRESS

Chapter 28: Show What You Know *continued*

Basic Cladogram

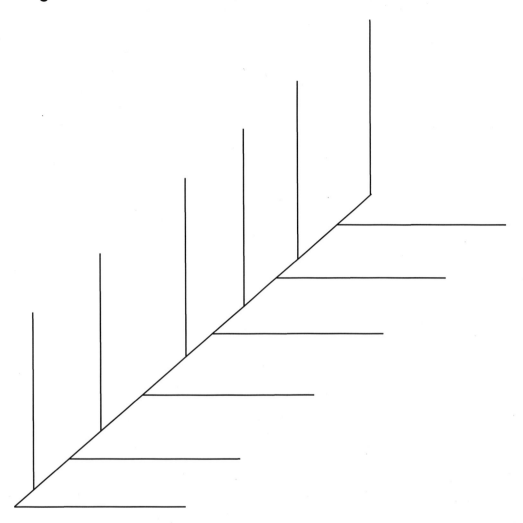

Organisms

1. **Pine tree**
2. **Fern**
3. **Macrocystis** (giant kelp that lives in the ocean)
4. **Fruit tree**
5. **Moss**
6. **Green algae** (a freshwater unicellular organism)

Shared Traits

1. **Flower**
2. **Vascular tissue**, xylem and phloem
3. **Seeds**
4. **Multicellular**
5. **Can live on land**
6. **Can photosynthesize**

Chapter 28: Show What You Know *continued*

Advanced Cladogram

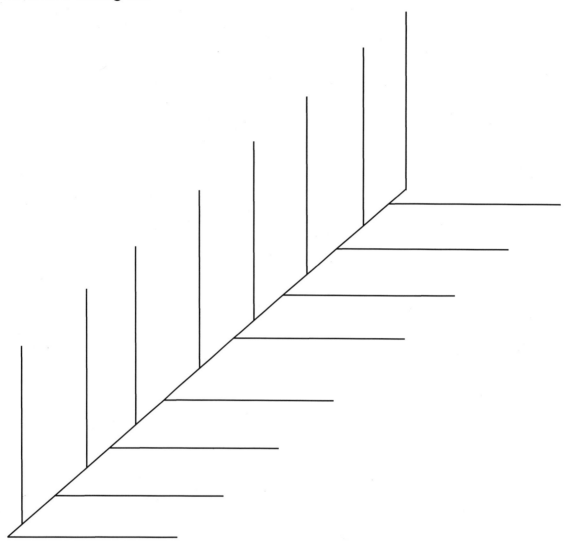

Organisms
1. Walrus
2. Honeybee
3. Rabbit
4. Fish
5. House cat
6. Pine tree
7. Turtle
8. Wolf

Shared Traits
1. Heterotroph
2. Lungs
3. Pointed teeth
4. Fur
5. Retractable claws
6. Backbone
7. Multicellular
8. Eyes facing forward (on front of head, the eyes of a predator)

Chapter 29: Domains Bacteria and Archaea

The Good Guys

Chapter 29: Microscope Lab

Bacteria are kind of like people, you have your good guys and your bad guys. Bad bacteria, pathogens, can make a person sick and in extreme cases even kill them. On the other hand, *without* good bacteria inside us, we would die! Good bacteria are bacteria that are beneficial for your health. Your intestines are full of good bacteria (and good archaea too) that are essential for your body to digest food. Good bacteria also live on your skin and in your mouth, protecting you from pathogens. When these bacteria encounter pathogens, they secrete chemicals that fight them.

Yogurt is full of good bacteria. Today you will use your microscope to look at the good bacteria in the yogurt. Then you will put the yogurt in a warm, dry place and look at it 24 hours later. What do you think will be different about it?

Materials

- Plain, active-culture yogurt
- Microscope
- Toothpick
- Water
- Dropper
- Slide and slide cover
- Methylene blue

Procedure

Day 1 1. Use the toothpick to smear a small sample of yogurt onto the slide. Put a drop of water on the slide and place a slide cover on top. Wait 15 minutes for the bacteria to settle on the slide.

2. Use 100x magnification to locate a sample with a good number of bacteria. Then switch to the 400x magnification to find a sample to sketch. Draw what you see on the lab sheet and answer the questions for Day 1. Next, stain the slide for a view with better definition of organism, but without movement. Stain kills bacteria.

3. Put the yogurt container in a dark and fairly warm location. Leave the container for 24 hours.

Day 2 4. Use the toothpick to smear a small sample of yogurt onto the slide. (Try to smear the same amount as you did yesterday.) Put a drop of water on the slide and place a slide cover on top.

5. Use 100x magnification to locate a sample with a good number of bacteria. Then switch to the 400x magnification to find a sample to sketch. Draw what you see on the lab sheet and answer the questions for Day 2. Next, stain the slide to view again.

The Good Guys

Chapter 29: Microscope Lab Sheet

Name_____ **Date**_____

Specimen _____ Type of mount_____

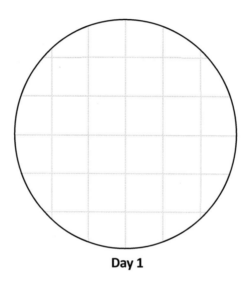

Day 1

Day 1

Read the label of the yogurt container. Write the names of the types of bacteria in the list of ingredients.

The bacteria in yogurt can be recognized by their shape or grouping arrangement. Search the names of the bacteria on your computer, ask, "What shape are *(bacteria name)*?" Record the types of bacteria you see with your microscope. Which bacteria are most abundant? Which bacteria are least abundant?

Chapter 29: Microscope Lab Sheet *continued*

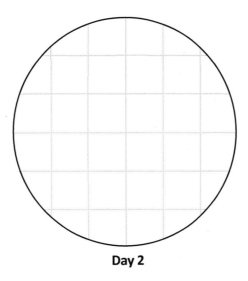

Day 2

Day 2

Did you see anything new? Did the abundance of any type of bacteria change more than others? Give explanations with your answers.

We have a symbiotic relationship with the good bacteria found in yogurt. These bacteria help us digest food and fight pathogens that get inside us. What benefit do we provide to the bacteria?

The Archaea of Yellowstone
Chapter 29: Famous Science Series

Famous Extremophiles

Locate Yellowstone National Park on a map of the United States.

Research archaea found in Yellowstone National Park on your computer. What colors are the mats of archaea?

What is the temperature of the water leaving the ground at the geysers in Yellowstone?

Do archaea live in the hot water around these geysers?

At what temperature does water boil?

The American Burn Association recommends your bathwater be what temperature?

Research information about the temperatures at the thermal pools in Yellowstone National Park. Could you live in the water of these pools?

Domains Bacteria and Archaea
Chapter 29: Show What You Know

Questions

1. The words outside the box are out of order. Write the words outside the box in the correct places in the table on the left. This is the classification for the pathogenic bacteria that is the leading cause of bacterial pneumonia.

Domain	
	Bacteria
	Firmicutes
	Bacilli
	Lactobacillales
	Streptococcaceae
Genus	
Species	

Order

Streptococcus

pneumoniae

Phylum

Class

Kingdom

Family

Bacteria

2. What shape do you think streptococcus bacteria have? Do they occur in groups of two, clusters, or in chains?

3. All cells have ribosomes, whether they are bacteria, archaea, or eukarya. Why? Hint: What do ribosomes do that is so important?

Chapter 29: Show What You Know *continued*

Multiple Choice

1. Why was the new classification of archaea made?

   ○ There were too many species in the group bacteria.
   ○ Archaea have eukaryotic cells.
   ○ Archaea use very different materials to build cellular structures.
   ○ Because of cladistics

2. Bacteria are responsible for

   ○ helping digest foods.
   ○ cycling nutrients.
   ○ some diseases.
   ○ All of the above

3. What shape are lactobacillus bacteria?

   ○ Spherical
   ○ Rod-shaped
   ○ Spiral
   ○ Clustered

4. A pathogen is

   ○ disease-causing.
   ○ a decomposer.
   ○ benign.
   ○ beneficial.

5. Which of the choices is NOT a trait of bacteria?

   ○ They are unicellular.
   ○ They reproduce asexually.
   ○ They are eukaryotic.
   ○ They have no membrane-bound nucleus.

6. Bacteria are classified by

   ○ feeding strategy.
   ○ shape.
   ○ the conditions they grow in.
   ○ genetic differences.
   ○ All of the above

7. Bacteria use their flagellum to

   ○ kill predators.
   ○ catch food.
   ○ move.
   ○ absorb nutrients.

8. The DNA of archaea and bacteria is

   ○ in a nucleus.
   ○ in a circular ring.
   ○ very similar.
   ○ They don't have DNA; they have RNA instead.

9. Archaea are classified using

   ○ feeding strategy.
   ○ shape.
   ○ the conditions they grow in.
   ○ differences in their RNA.
   ○ All of the above

10. Cyanobacteria are autotrophs because

    ○ they have chlorophyll in their cell membranes, which they use for photosynthesis.
    ○ they eat other bacteria.
    ○ they absorb food through their cell membrane.
    ○ they are anaerobic.

Chapter 29: Show What You Know *continued*

11. Which of these is NOT true of archaea?

    ○ They are unicellular.
    ○ They reproduce asexually.
    ○ They are prokaryotic.
    ○ They have RNA but no DNA.

12. Anaerobic organisms

    ○ feed themselves by absorbing nutrients.
    ○ cannot survive in the presence of atmospheric oxygen.
    ○ are aquatic.
    ○ need atmospheric oxygen.

13. You know from the name that spirillum bacteria are

    ○ aerobic.
    ○ anaerobic.
    ○ green.
    ○ spiral.

14. Pili help bacteria

    ○ attach to surfaces.
    ○ move.
    ○ defend themselves.
    ○ absorb nutrients.

15. Archaea are sometimes called extremophiles because they

    ○ are extremely hard to culture.
    ○ live in extreme environments that would kill other organisms.
    ○ are extremely rare.
    ○ are extremely sensitive to the presence of atmospheric oxygen.

Chapter 30: Kingdom Plantae

Perusing Plants
Chapter 30: Lab

Now that you know how plants are classified, can you identify the different types? You are about to find out if you can or can't. For this lab, you will look around your house, outside, and possibly take a "field" trip to a nursery. You will try to find all the divisions of plants. You will also identify two of the classes, monocots and dicots.

Materials

- Field guide for plants (optional)
- Outdoor plants to observe

Procedure

1. If you have plants in your house, try to determine their division. If the plant is an angiosperm, determine whether it is a monocot or a dicot.

2. Go outside and start identifying plants. Try to find multiple examples of each division. When you identify an angiosperm, determine if it is a monocot or a dicot. See if you can find a plant for every division and class that is discussed. Use your field guide to help if you get stuck.

Perusing Plants

Chapter 30: Lab Sheet

Name_____ **Date**_____

Keep track of each type of plant as you determine what it is. This does not have to be detailed. You can draw a quick sketch of the plant, write its species name (if you know it), or even just use a hash mark. Do not repeat the same species of plant when you are counting, but you should keep track of the number of different species you find in each division and class.

Bryophytes (moss—bryophytes live in moist environments)

Pterophytes (ferns)

Gymnosperms (cycads, ginkgoes, conifers)

Angiosperms

Monocots (have long leaves with parallel veins, and they have petals in multiples of 3)

Dicots (have webbed veins, and they have petals in multiples of 4 or 5)

Pandia PRESS

Distinguishing Dicots and Monocots
Chapter 30: Microscope Lab

One of the distinguishing traits for a monocot and a dicot is their leaf structure. Do you think you will be able to see a difference with your microscope?

Materials

- Blade of grass (monocot leaf)
- Leaf from a tree that is an angiosperm (dicot leaf), any tree except a palm tree or a bamboo plant

- Scalpel or paring knife
- 2 slides
- 2 slide covers

- Water
- Syringe
- Microscope

Procedure

1. Cut a square piece of each leaf and make a wet mount slide for each of them.

2. Compare the two types of leaf.

3. Draw a view of each at your favorite magnification for them. Make sure that you choose the same magnification for both leaves. Pay special attention to the differences between the two leaves.

Distinguishing Dicots and Monocots

Chapter 30: Microscope Lab Sheet

Name_____ **Date**_____

Specimen _____ Type of mount_____

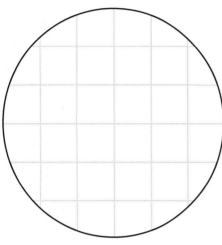

Monocot leaf (blade of grass)

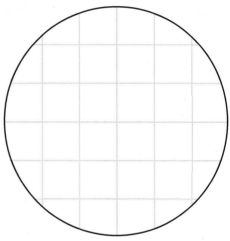

Dicot leaf (tree leaf)

Observations

If you know the name of the type of grass or tree, write it here.

Describe the difference in the way the veins look in both leaves.

Pandia PRESS

George Washington Carver

Chapter 30: Famous Science Series

Famous Botanist

When and where was George Washington Carver born? What was the situation of his birth?

George Washington Carver

Carver received a bachelor's degree in agriculture in 1894. Where was it from?

While he was studying for his master's, Carver was approached by another famous African-American man about teaching at the college where he was principal. Who was the man and what was the name of the college?

Carver was a researcher as well as a teacher. He studied plants. What plants did he study?

Carver invented 300 uses for one of these. What was it?

George Washington Carver developed crop-rotation methods that helped farmers in the South. What is crop rotation and how does it help farmers? What two crops did Carver rotate and why did this help?

Kingdom Plantae
Chapter 30: Show What You Know

Questions

1. The oldest known living thing on Earth is a little less than 5,000 years old. It is a bristlecone pine tree named Methuselah. Methuselah lives in the Ancient Bristlecone Pine Forest in the White Mountains in California. Write the classification for a species of bristlecone pine in the chart below. Write the words outside the box in the correct places in the table.

	Eukarya	
	Plantae	
	Pinophyta	
	Pinopsida	
	Pinales	
	Pinaceae	
Genus		
Species		

Pinus

Class

Domain

Order

longaeva

Division

Family

Kingdom

What is the scientific name for this species of bristlecone pine?

2. Fill in the cladogram. Put the shared derived traits on the horizontal lines and the organisms at the top of the vertical lines.

Organisms
1. Fern
2. Pine tree
3. Moss
4. Rose bush

Shared Derived Traits
1. Seeds
2. Can photosynthesize
3. Flowers
4. Vascular tissue

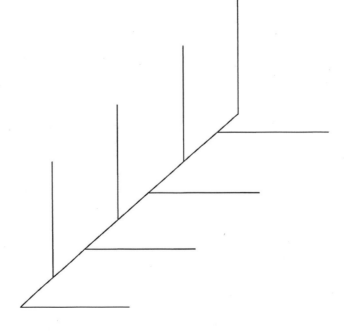

Chapter 30: Show What You Know *continued*

3. Each one of the statements about plants is false. Fix each statement to make it true.

All plants

. . . are unicellular. _____

. . . are mobile. _____

. . . have prokaryotic cells. _____

. . . have cell walls made from glucose. _____

. . . are heterotrophs, which use cellular respiration to make food. _____
_____

. . . reproduce using binary fission. _____

Multiple Choice

1. You are walking along a creek and you see a clump of low-growing plants. When you examine the clump more closely, you see it has no roots or leaves. It is (a)

   ○ moss.
   ○ fern.
   ○ cycad.
   ○ daisy.

2. You walk further and see a plant with an interesting leaf. You turn the leaf over and see rows of button-like spores running down it. It is a

   ○ moss.
   ○ fern.
   ○ cycad.
   ○ daisy.

3. Next, you reach down and pick a flower. It is a(n)

   ○ bryophyte.
   ○ gymnosperm.
   ○ angiosperm.
   ○ pterophyte.

4. The flower has 9 petals; it is a

   ○ pterophyte.
   ○ bryophyte.
   ○ monocot.
   ○ dicot.

5. Another name for a bryophyte is a

   ○ monocot.
   ○ dicot.
   ○ moss.
   ○ fern.

6. Another name for a pterophyte is a

   ○ monocot.
   ○ dicot.
   ○ moss.
   ○ fern.

7. Seeds

   ○ have a food source enclosed inside with the embryo.
   ○ are transported in the poop of animals like bears and deposited far away from their parent plant.
   ○ have a coating that protects them.
   ○ All of the above

Chapter 30: Show What You Know *continued*

8. A plant with no vascular tissue is a(n)

   ○ bryophyte.
   ○ pterophyte.
   ○ angiosperm.
   ○ gymnosperm.

9. A plant with vascular tissue that uses spores to reproduce is a(n)

   ○ bryophyte.
   ○ pterophyte.
   ○ angiosperm.
   ○ gymnosperm.

10. A plant that uses seeds arranged on cones to reproduce is a(n)

   ○ bryophyte.
   ○ pterophyte.
   ○ angiosperm.
   ○ gymnosperm.

11. A plant that makes flowers is a(n)

   ○ bryophyte.
   ○ pterophyte.
   ○ angiosperm.
   ○ gymnosperm.

12. A plant's vascular system

   ○ is used for reproduction.
   ○ transports materials to all parts of the plant.
   ○ makes glucose.
   ○ is only found in bryophytes.

13. If a daisy has 88 petals, it is a

   ○ monocot.
   ○ dicot.
   ○ bryophyte.
   ○ pterophyte.

14. Bamboo has long, parallel veins in its leaves; it is a

   ○ monocot.
   ○ dicot.
   ○ bryophyte.
   ○ pterophyte.

15. Spores

   ○ have been important for plants colonizing harsh environments.
   ○ have a food store, so they can go a long time before germinating.
   ○ need water for part of their life cycle.
   ○ allowed gymnosperms to be the dominant land plant 100 million years ago.

Chapter 31: Kingdom Animalia

Arthropod Arrangement

Chapter 31: Lab and Microscope Lab

Insects are in class Hexapoda. Spiders and ticks are in class Arachnida.

If you were asked to list the differences between mammals and insects, it would be easy, wouldn't it? The more related two organisms are, the harder it is, though. Today you are going to examine an arachnid and a hexapod. Arachnids are spiders or ticks. Hexapods are insects. Because they are both arthropods, they do have many things in common. They share more derived traits with each other than either shares with mammals. They are two different classes of arthropods, though, so there are differences too. What do you think the differences are? Today you will research and find out!

No arthropods have to be harmed to perform this experiment! Take a few days leading up to this experiment looking on your windowsills or outside on the ground for dead spiders and insects; maybe you can find a dead (hopefully) tick on your dog or cat. You will probably find many parts and pieces of specimens, but you need at least one good arachnid and one good insect that are undamaged and are completely intact. That smushed mosquito on your arm might be cool, but it's not going to be a good anatomical specimen for under your microscope

Materials

- Magnifying glass
- Arachnid specimen (spider)
- Insect specimen
- Microscope (optional)
- White-out or liquid paper

- Scalpel
- Slides (optional)
- Slide covers (optional)
- A white sheet of copy paper
- Syringe

- Water
- Tweezers
- Flashlight

Chapter 31: Lab and Microscope Lab *continued*

Procedure

Be VERY careful when you are working with the specimens; they are delicate.

1. Put the spider and the insect next to each other on the piece of white paper. Examine them with your naked eye and with the magnifying glass. Be careful not to damage them as you examine them, but try to move them around and look at them from all sides; the tweezers and flashlight can be useful for this. As you examine them, find the labeled parts on your lab sheet.

2. The illustrations are not very detailed. You can add details to the drawings if you like. You might need to remove details as well. For instance, not all insects have wings. White-out any body parts missing from your insect. Count the arachnid's and insect's legs; do they have the correct number?

3. All arthropods have an exoskeleton, jointed legs, and multiple body parts.

   • Gently, without smashing them, examine the exoskeletons with your naked eye, the magnifying glass, and the microscope (optional). On your lab sheet describe what you see.

   • How do the body parts connect? Examine the site of connection. On your lab sheet describe what you see.

4. Take the specimens off and carefully begin dissecting each. For example, use the scalpel and tweezers and look at the leg from one of the animals and then the leg from the other.

   • Look at the leg joints on both animals. How are they similar? On your lab sheet describe what you see.

5. On your lab sheet, summarize the differences and similarities between hexapods and arachnids, using your insect and your spider as examples.

There is a weakness with this experiment; have you thought of what it is? This experiment assumes the one insect and the one arachnid represent all hexapods and all arachnids. The scientists that classified arachnids and insects and divided them into different classes looked at many different species of arachnids and insects, not just one of each class.

Explore

Arthropod Arrangement

Chapter 31: Lab Sheet

Name_____ **Date**_____

My Insect, Class Hexapoda

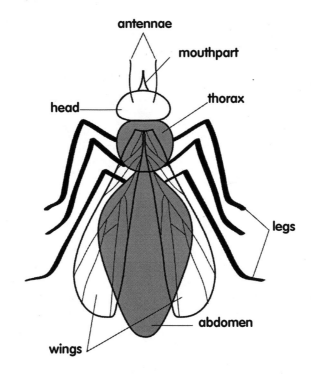

antennae

mouthpart

head

thorax

legs

abdomen

wings

My Observations

Exoskeleton:

Body parts:

Joints and legs:

Other observations:

Chapter 31: Lab Sheet *continued*

My Arachnid, Class Arachnida

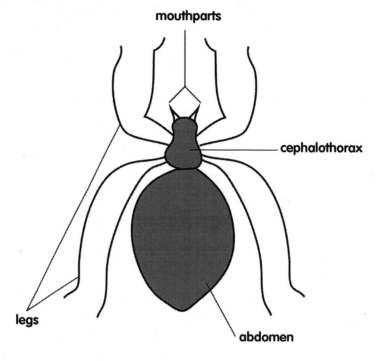

My Observations

Exoskeleton:

Body parts:

Joints and legs:

Other observations:

Similarities and Differences of Hexapods and Arachnids

Pandia PRESS

John James Audubon

Chapter 31: Famous Science Series

Famous Observer of Birds

John James Audubon painted hundreds of birds. Where was he born? When was he born? In what country did he study birds? Why did he go to that country?

James Audubon

What was the name of the book Audubon wrote about birds?

What society is named after him? What is its purpose?

When Audubon discovered a bird he had not seen before, he shot it to study it more closely. Then he painted it. Do you think members of the National Audubon Society still shoot birds to study them?

Kingdom Animalia
Chapter 31: Show What You Know

1. The words outside the box are out of order. Write those words in the correct places in the table on the left. This is the classification for the Southern Hairy-Nosed Wombat. Wombats are marsupials native to Australia and the island of Tasmania.

	Eukarya
	Animalia
	Chordata
	Mammalia
	Diprotodontia
	Vombatidae
	Lasiorhinus
	latifrons

Order

Kingdom

Genus

Phylum

Class

Domain

Species

Family

What is the scientific name for the Southern Hairy-Nosed Wombat? _____

2. Each one of the statements about animals is false. Fix each statement to make it true.

All animals . . .

are unicellular. _____

are immobile. _____

have prokaryotic cells. _____

have cell walls. _____

are autotrophs, which means they make their own food using photosynthesis. _____

_____

3. There are four groups of arthropods. Match each group with the best description for it.

Crustacea   long bodies, lots of segments, 1 to 2 feet coming from each segment

Myriapoda   2 body segments, 8 legs

Hexapoda   3 body segments, 6 legs, 2 antennae

Arachnida   10 to 40 legs, 4 antennae, gills

Pandia PRESS

Chapter 31: Show What You Know *continued*

4. Fill in the cladogram. Put the shared derived traits on the horizontal lines and the organisms at the top of the vertical lines.

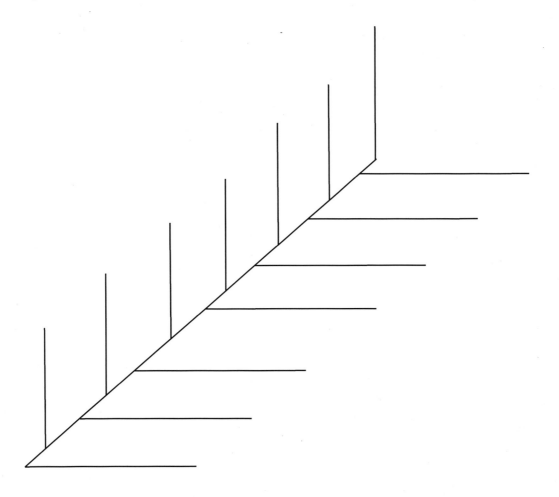

Organisms

1. Opossum

2. Army ant

3. Starfish

4. Duckbilled platypus

5. Red-tail hawk

6. Mountain gorilla

7. Chameleon

Shared Traits

1. Heterotroph

2. Placenta develops in female when she is pregnant

3. Endoskeleton

4. Endotherm

5. Lives on land

6. Mammary glands

7. Long pregnancy, followed by the delivery of more developed offspring

Chapter 31: Show What You Know *continued*

Multiple Choice

1. A squid is an invertebrate animal, meaning

   ○ it lives in water.
   ○ it has a backbone.
   ○ it does not have a backbone.
   ○ it is cold-blooded.

2. An animal that has one muscular foot and a soft body is a(n)

   ○ mollusk.
   ○ echinoderm.
   ○ porifera.
   ○ arthropod.
   ○ cnidaria.
   ○ chordate.

3. An animal that is aquatic, does not have tissue but does have specialized cells, has a hollow body with pores in it, and has a big hole on top where waste flows out, is a(n)

   ○ mollusk.
   ○ echinoderm.
   ○ porifera.
   ○ arthropod.
   ○ cnidaria.
   ○ chordate.

4. An animal with jointed legs, a segmented body, and an exoskeleton is a(n)

   ○ mollusk.
   ○ echinoderm.
   ○ porifera.
   ○ arthropod.
   ○ cnidaria.
   ○ chordate.

5. An aquatic animal with a radial body plan, a sac-like body with one opening, and stinging tentacles it uses to immobilize it prey is a(n)

   ○ mollusk.
   ○ echinoderm.
   ○ porifera.
   ○ arthropod.
   ○ cnidaria.
   ○ chordate.

6. An aquatic animal with a radial body plan, a tough spiny skin, and tube feet it uses to move is a(n)

   ○ mollusk.
   ○ echinoderm.
   ○ porifera.
   ○ arthropod.
   ○ cnidaria.
   ○ chordate.

7. An animal with a backbone, a head, and a sophisticated body plan is a(n)

   ○ mollusk.
   ○ echinoderm.
   ○ porifera.
   ○ arthropod.
   ○ cnidaria.
   ○ chordate.

8. Segmented worms

   ○ have both male and female parts.
   ○ are endotherms.
   ○ have a flat body with a mouth at one end.
   ○ have a long, threadlike body.

Chapter 31: Show What You Know *continued*

9. Nematodes are roundworms that

   ○ have segmented, tube-shaped bodies.

   ○ are endotherms.

   ○ have a flat body with a mouth at one end.

   ○ have a long, threadlike body.

10. Platyheminthes are worms that

    ○ have segmented, tube-shaped bodies.

    ○ are endotherms.

    ○ have a flat body with a mouth at one end.

    ○ have a long, threadlike body.

11. This animal gets food when it moves food and water through pores in its body. It is a

    ○ starfish.

    ○ worm.

    ○ sea anemone.

    ○ sponge.

12. The term *radial body plan* means an organism

    ○ has a backbone.

    ○ has no backbone.

    ○ has a central point that the rest of their body is arranged around.

    ○ has a sac-like body in the shape of a circle.

13. A vertebrate has

    ○ an exoskeleton.

    ○ an internal skeleton.

    ○ a hard coating on the outside of its body.

    ○ a shell.

14. Endoskeletons are made from

    ○ bone and cartilage.

    ○ chitin.

    ○ collagen.

    ○ glycogen.

15. An ectotherm

    ○ regulates its own body temperature internally.

    ○ regulates its body temperature by exchanging heat with the environment.

    ○ has no endoskeleton.

    ○ has no backbone.

16. An endotherm

    ○ regulates its own body temperature internally.

    ○ regulates its body temperature by exchanging heat with the environment.

    ○ has no endoskeleton.

    ○ has no backbone.

17. This vertebrate animal goes through a metamorphosis, where it starts out as one form and grows to look differently as an adult, is an ectotherm, and lays eggs in water. It is a(n)

    ○ mamma.

    ○ reptile.

    ○ amphibian.

    ○ bird.

    ○ fish.

Chapter 31: Show What You Know *continued*

18. This vertebrate animal lives in water, has fins, breathes through gills, lays eggs, and is an ectotherm. It is a(n)

    ○ mammal.
    ○ reptile.
    ○ amphibian.
    ○ bird.
    ○ fish.

19. This vertebrate animal has feathers and wings, a beak, lays eggs, and is an endotherm. It is a(n)

    ○ mammal.
    ○ reptile.
    ○ amphibian.
    ○ bird.
    ○ fish.

20. This vertebrate animal has dry scaly skin, lays eggs, and is an ectotherm. It is a(n)

    ○ mammal.
    ○ reptile.
    ○ amphibian.
    ○ bird.
    ○ fish.

21. This vertebrate animal has mammary glands, hair or fur, and is an endotherm. It is a(n)

    ○ mammal.
    ○ reptile.
    ○ amphibian.
    ○ bird.
    ○ fish.

22. The purpose of mammary glands is to

    ○ regulate body temperature.
    ○ make hormones.
    ○ make milk.
    ○ fight infections.

23. A mammal that lays eggs is a

    ○ marsupial.
    ○ monotreme.
    ○ placental mammal.
    ○ mammals do not lay eggs.

24. Organisms with gills

    ○ transfer oxygen and carbon dioxide across them.
    ○ are only fish.
    ○ are anaerobic.
    ○ use sulfur dioxide for cellular respiration.

Pandia PRESS

Chapter 32: Kingdoms Fungi and Protists

Watching Fungi Feed

Chapter 32: Lab

This experiment takes 5 to 10 days from start to finish. The number of days depends on the ripeness of the banana and the temperature where the experiment is conducted.

You know how autotrophs, such as plants, make food. You know how heterotrophs, such as yourself, eat. You even know how heterotrophs, like amoebas, eat using endocytosis. But what about heterotrophs that are decomposers? How does that work? What does it look like?

Yeast are unicellular fungi. They are chemotrophs like all fungi. They secrete chemicals that turn their food into molecules they can absorb. This decomposes food. You will look at how yeast "eat" a banana. Bananas have glucose in them. Yeast break down the glucose molecules to make food they can absorb. When they do that, carbon dioxide, a gas, is released. Do you think fungi are fast or slow eaters?

You will divide the banana in half. You will put one of the halves in a baggie with the yeast. The other banana will be put in an empty baggie. The untreated banana is a control. A ***control*** in an experiment is used as a standard for comparison. The only difference between the two groups will be that one half of the banana was treated with yeast and the other half was not. All other parameters are kept the same. Therefore, any differences can be attributed to the action of the yeast on the banana.

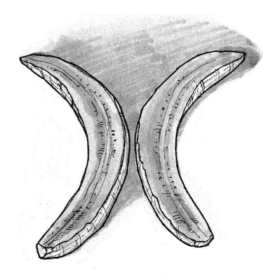

Chapter 32: Lab *continued*

Materials

- 1 very ripe banana
- 2 zipper-lock baggies

- 1 teaspoon of yeast
- String

- Ruler

Procedure

1. Write your hypothesis on the lab sheet. What do you think will happen to the banana in the baggie with the yeast?

Day 1

2. Peel the banana and slice it in half lengthwise. Put each half in a separate baggie. Sprinkle yeast into one of the baggies. Shake it around to make sure the banana is well coated in yeast. Get as much air as possible out of both baggies before sealing them.

3. Write the first day's observations; you can use words and/or drawings.

Days 2 to 4 (or more, until you see results)

4. Look at both baggies. Do not open either baggie before the final day. Write down that day's observations. If one baggie is bigger than the other, measure it around with the string, then measure the string with the ruler.

Final Day

5. Make a final measure of the baggie if it has been getting bigger. Then open both baggies and record your observations. Did carbon dioxide accumulate in one of the baggies? Did you observe the yeast decomposing the banana?

6. Write the results of the experiment. What is the difference between the baggie with the yeast in it and the control baggie? A drawing or a photograph should accompany your written description of the two banana halves. Did the results match your hypothesis? If yes, then your conclusion would be the same as your hypothesis. If your hypothesis was wrong or only partially correct, write a new conclusion.

Watching Fungi Feed

Chapter 32: Lab Sheet

Name_____ Date_____

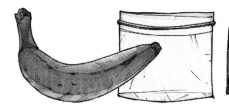

Hypothesis:

Observations

Day 1:

Day 2:

Day 3:

Day 4:

Day 5:

Chapter 32: Lab Sheet *continued*

Results/Conclusion:

Bonus question: What reaction do you think is occurring if carbon dioxide is accumulating inside the baggie? Why is it accumulating and not cycling? What would happen if the gas inside the baggie ran out of oxygen?

Pandia PRESS

Explore Fungus Up Close and Personal
Chapter 32: Dissection and Microscope Lab

Materials

- Whole mushroom (includes cap and stem)*
- Scalpel
- Slides
- Slide covers
- Microscope
- Cutting board
- Water

- Methylene blue
- Syringe
- Forceps
- Parts of a mushroom diagram from the lesson in the textbook
- Flashlight. You will need top lighting if your slices are thick.

*Find a whole mushroom in your grocery store or farmers market, preferably one that has what look like roots (hyphae) at the bottom.

Procedure

1. Examine the mushroom. Identify the parts of it. Use the diagram from your textbook if you need help. You might need to remove a layer of tissue, called a veil, covering the gills. If you need to remove a veil, do it carefully without touching the gills.

2. Very gently, separate the cap from the stem by twisting it. Carefully set the cap aside gillside up. Draw the underside of the cap in the box entitled My Mushroom. Label the hyphae, gills, and cap in your drawing.

3. Now examine the cap. The gills are lined with structures called basidia. Basidia are where spores are produced on mushrooms. They are small and fragile; be gentle. With your scalpel, slice one of the gills from the cap. DO NOT touch the free edge if you can avoid it. Using methylene blue, make a stained wet mount slide with the gill. Draw a microscope view of the gill, basidia, and spores at your favorite magnification.

4. Turn the cap over and make a thin slice of the mushroom cap. If the cap has scales, try to include them in your slice. Using methylene blue, make a stained wet mount slide of the cap. Draw a microscope view of the cap at your favorite magnification.

Fungus Up Close and Personal

Chapter 32: Microscope Lab Sheet

Name_____ **Date**_____

Specimen _____ Type of mount_____

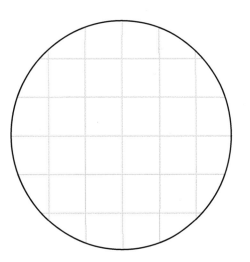

My Mushroom

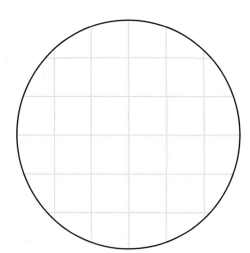

Hyphae magnification _____

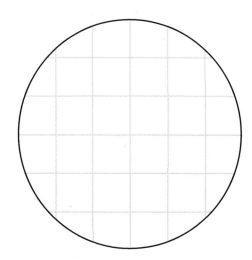

Gill, basidia, and spores magnification _____

Cap magnification _____

Did you see any chloroplasts? Why or why not?

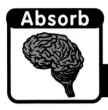

Absorb

Truffle Pigs

Chapter 32: Famous Science Series

Famous Fungus Finders

What are truffles, what are they used for, and what is the job of a truffle pig?

Where do truffles grow?

When did people start using pigs to locate truffles?

Another animal is replacing the truffle pig. What kind of animal is it and why is the pig being replaced?

Kingdoms Fungi and Protists
Chapter 32: Show What You Know

Cladogram

Fill in the cladogram. Put the shared derived traits on the horizontal lines and the organisms at the top of the vertical lines.

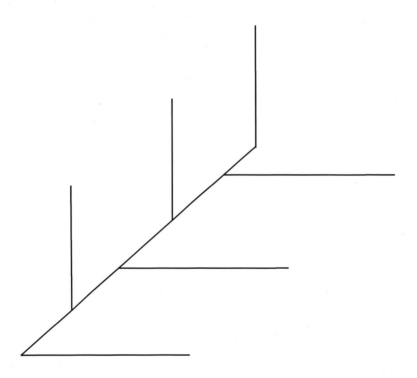

Organisms	Shared Traits
1. amoeba	1. chemotroph
2. mushroom	2. is not mobile
3. slime mold	3. eukaryote

Pandia PRESS

Chapter 32: Show What You Know *continued*

Classification

The words outsode the box are out of order. Write those words in the correct places in the table on the left. This is the classification for a species of truffle.

Domain	
Kingdom	
	Ascomycota
	Pezizomycetes
	Pezizales
	Tuberaceae
Genus	
Species	

Tuber

Order

Fungi

Phylum

borchii

Eukarya

Class

Family

What is the scientific name of this truffle?

Multiple Choice

1. Algae are

   ○ plantlike protists.
   ○ animal-like protists.
   ○ fungi-like protists.
   ○ fungi.

2. Amoebas are

   ○ plantlike protists.
   ○ animal-like protists.
   ○ fungi-like protists.
   ○ fungi.

3. Mushrooms are

   ○ plantlike protists.
   ○ animal-like protists.
   ○ fungi-like protists.
   ○ fungi.

4. Slime molds are

   ○ plantlike protists.
   ○ animal-like protists.
   ○ fungi-like protists.
   ○ fungi.

5. Plantlike protists are

   ○ heterotrophs.
   ○ autotrophs.
   ○ chemotrophs.
   ○ All of the above
   ○ None of the above

6. Animal-like protists are

   ○ heterotrophs.
   ○ autotrophs.
   ○ chemotrophs.
   ○ All of the above
   ○ None of the above

Chapter 32: Show What You Know *continued*

7. Fungi-like protists are

   ○ heterotrophs.
   ○ chemotrophs.
   ○ All of the above
   ○ None of the above

8. Which characteristic is NOT shared between fungi and fungi-like protists?

   ○ They are eukaryotes.
   ○ They are chemotrophs.
   ○ They are mobile.
   ○ They reproduce by forming spores.

9. Which characteristic is NOT shared between animals and animal-like protists?

   ○ They are eukaryotes.
   ○ They are heterotrophs.
   ○ They are mobile.
   ○ They are unicellular.

10. Which characteristic is NOT shared between plants and plant-like protists?

   ○ They are eukaryotes.
   ○ They are autotrophs.
   ○ They are mobile.
   ○ They photosynthesize.

11. Protists

   ○ are eukaryotes.
   ○ are usually unicellular.
   ○ live in watery environments.
   ○ All of the above

12. Chemotrophs

   ○ make glucose through a chemical process.
   ○ get their energy from the chemicals in plants and animals.
   ○ secrete chemicals that break material into molecules they can absorb.
   ○ feed themselves using the chemical process called exocytosis.

13. Lichens are a symbiotic relationship between what two types of organisms?

   ○ Fungi and algae
   ○ Fungi and amoeba
   ○ Fungi and slime mold
   ○ Algae and amoeba
   ○ Amoeba and slime mold
   ○ Algae and slime mold

14. Fungi reproduce

   ○ asexually.
   ○ using spores.
   ○ by binary fission.
   ○ without making gametes.

15. Hyphae

   ○ are the reproductive organ of fungi.
   ○ are used for absorbing nutrients.
   ○ are photosynthetic.
   ○ are unicellular.

Test

Appendix A

Unit Exams

Units I and II: Organisms and Cells
Exam Chapters 1 - 6

1. Multiple Choice

The organelle where photosynthesis occurs is the

- ○ mitochondria
- ○ chloroplast
- ○ nucleus
- ○ vacuole

An organism whose DNA is in a nucleus is a

- ○ virus
- ○ prokaryote
- ○ eukaryote
- ○ bacteria

The six main elements organisms are made from are

- ○ hydrogen, sulfur, lead, oxygen, carbon, sodium
- ○ hydrogen, oxygen, carbon, nitrogen, phosphorus, calcium
- ○ nitrogen, phosphorus, calcium, silicon, sulfur, helium
- ○ lithium, beryllium, boron, carbon, nitrogen, oxygen

The building block of all living things is the

- ○ cell
- ○ atom
- ○ carbon
- ○ matter

Semi-permeable membranes

- ○ are not found in organisms
- ○ let all things pass through
- ○ let some but not all things pass through
- ○ do not allow material to diffuse across them

This organelle gives plants the structure they need to stand tall.

- ○ Chloroplast
- ○ Nucleus
- ○ Cell membrane
- ○ Cell wall

Because the cells of multicellular organisms specialize,

- ○ they all are the same size and shape
- ○ one cell from a multicellular organism could survive on its own
- ○ there are many different sizes and shapes
- ○ they are smaller than the cells of unicellular organisms

Fermentation

- ○ is a type of respiration
- ○ occurs when no oxygen is present
- ○ generates less energy than cellular respiration
- ○ occurs in the cytoplasm
- ○ all of the above
- ○ none of the above

Passive transport

- ○ does not use energy
- ○ is an example of endocytosis
- ○ moves molecules from an area of low to high concentration
- ○ does use energy

The cell theory states that

- ○ cells come only from other living cells
- ○ all organisms are made from one or more cells
- ○ cells carry out all the functions needed for life
- ○ all of the above

Some scientists think viruses should be reclassified as organisms because they

- ○ are able to grow
- ○ are able to reproduce
- ○ take in energy
- ○ are made of cells

After a virus has infected a cell, it turns the cell into

- ○ a virus cell
- ○ bacteria
- ○ a pathogen
- ○ a virus-making factory

2. Vocabulary

Match the word with the definition that best fits.

Organism ◯   ◯ Transport large molecules into cell

Prokaryote ◯   ◯ Group of all the same atoms

Multicellular ◯   ◯ Green molecules able to capture energy from the sun

Atoms ◯   ◯ Many-celled

Element ◯   ◯ Transport large molecules out of cell

Chemical energy ◯   ◯ Living being

Osmosis ◯   ◯ Molecule cells use for energy

Active transport ◯   ◯ Unicellular organism with DNA in the cytoplasm

Endocytosis ◯   ◯ Diffusion of water

Exocytosis ◯   ◯ Energy stored in the bonds of molecules

Glucose ◯   ◯ Building blocks of matter

Chlorophyll ◯   ◯ Movement of molecules from low to high concentration, requires energy

3. Fill in the blanks

Proteins are made from _____ molecules.

Organisms eat _____ molecules for energy.

Chemical reactions in an organism happen in _____ .

_____ molecules make hair, skin, and hemoglobin.

4. Draw a cell membrane. Include a protein with a protein channel. Label all the parts. Show diffusion across the cell membrane.

5. Chemical Reactions. Write the chemical reactions for photosynthesis and cellular respiration. Names of the molecules can be in words or chemical symbols. Give the number of each type of molecule.

Pandia PRESS

6. Organelles. Fill in the names of the organelles and other material in the cell below. Use their location and description as a guide.

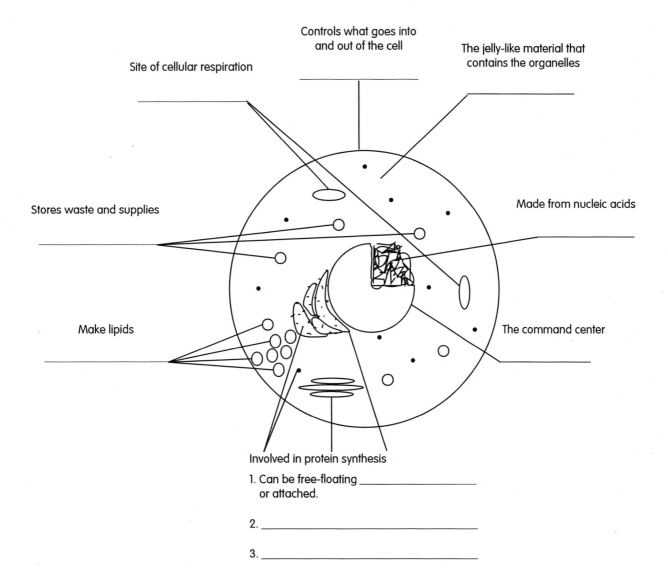

Controls what goes into and out of the cell

The jelly-like material that contains the organelles

Site of cellular respiration

Stores waste and supplies

Made from nucleic acids

Make lipids

The command center

Involved in protein synthesis

1. Can be free-floating _____ or attached.

2. _____

3. _____

What two organelles does a plant cell have that an animal cell doesn't?

1. _____

2. _____

7. Extra Credit

There are nine characteristics all organisms have in common. What are they?

1. _____

2. _____

3. _____

4. _____

5. _____

6. _____

7. _____

8. _____

9. _____

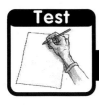

Unit III: Genetics
Exam Chapters 7–10

1. Vocabulary

Match the word with the definition that best fits.

Homologous chromosomes ◯   ◯ Division of cytoplasm, organelles, and cell

Zygote ◯   ◯ The complete set of genes in an organism's chromosomes

Gametes ◯   ◯ Using mitosis to make offspring

Haploid ◯   ◯ Diploid cell made from the process of fertilization

Genome ◯   ◯ A cell whose chromosome number is n

Asexual reproduction ◯   ◯ Chromosome pairs

Cytokinesis ◯   ◯ The process of copying DNA from a complementary strand of DNA

Replication ◯   ◯ Haploid sex cells

Allele ◯   ◯ Form of a gene

2. True or False (Extra points for correcting the false statements.)

_____ Cytokinesis is the part of the cell cycle when proteins are made.

_____ In humans, it is the gamete from the father that determines gender.

_____ During translation, RNA gives the information for building a protein to DNA.

_____ Meiosis results in genetically identical cells.

_____ DNA contains all the information needed to run your body.

_____ If an organism has 10 chromosomes in their somatic cells, their gametes will have 5 chromosomes in them.

_____ The genes you inherit are responsible for all your traits.

_____ The specific bonding of base pairs makes the replication of DNA strands possible.

_____ Meiosis starts with 1 cell and ends with 2 cells.

_____ A diploid cell has 2n chromosomes in it.

3. Multiple Choice

The acronym for remembering the steps mitosis and meiosis follow is

- ○ MAPT
- ○ MATP
- ○ PMAT
- ○ AMPT

The gender of humans is determined by the 23rd chromosome in the following pattern:

- ○ male XY, female XX
- ○ male XX, female XY
- ○ male XX, female XO
- ○ chromosomes don't determine gender

During prophase, the _____ become visible.

- ○ organelles
- ○ homologous chromosomes
- ○ genes
- ○ mitochondria

At the beginning of meiosis there is/are _____ chromosome(s). At the end of meiosis there are _____ chromosomes.

- ○ 2n, n
- ○ 1, 4
- ○ n, 2n
- ○ heterozygous, homologous

The three parts of the somatic cell cycle are

- ○ prophase, telophase, metaphase
- ○ mitosis, meiosis, cytokinesis
- ○ transcription, translation, ribosomes
- ○ interphase, mitosis, cytokinesis

DNA is in the shape of

- ○ a cone
- ○ a spheroid
- ○ a double helix
- ○ a tetrahedron

During fertilization, two gametes fuse to make a _____.

- ○ clone
- ○ zygote
- ○ haploid cell
- ○ bacterium

How many chromosomes are in a human skin cell?

- ○ 8
- ○ 23
- ○ 46
- ○ It depends if it is a male or a female

The correct order going from the smallest building block to the largest is

- ○ gene, codon, genome, nucleotide base, chromosome, homologous chromosome
- ○ homologous chromosome, chromosome, gene, genome, codon, nucleotide base
- ○ genome, homologous chromosome, chromosome, gene, codon, nucleotide base
- ○ nucleotide base, codon, gene, chromosome, homologous chromosome, genome

When bacteria reproduce using asexual reproduction, it is called

- ○ binary fission
- ○ budding
- ○ vegetative propagation
- ○ regeneration

4. Written Answers

Describe how a protein is made. Start your description of the events inside the nucleus. Make sure to use the correct vocabulary terms when describing the process of protein synthesis.

Draw a complementary strand of DNA above this sequence, and draw a complementary strand of RNA below this sequence.

A T C G G T T A G C T A G C C

What is the name of the process where DNA makes a copy along a strand?

List two things that are different between mitosis and meiosis.

List two things that are the same for mitosis and meiosis.

Put these in correct order using numbers from 1 to 4, 1 being the first stage of mitosis and 4 being the last stage of mitosis.

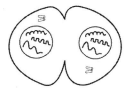 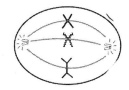 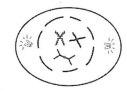 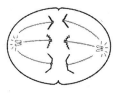

_____ _____ _____ _____

5. Punnett Square

You are so lucky. Your parents bought you two qwitekutesnutes for your birthday! The pet shop owner sold your mom two females. At least that's what he told her. Except that now it looks like one of them is going to have babies. You are really excited, and your mom is not. I wonder why? One of the qwitekutesnutes has 3 toes (tt) and one has 5 toes (Tt). If your qwitekutesnutes have 4 babies, how many should have 3 toes? How many should have 5 toes? To answer these questions, fill in the Punnett square. The allele for 3 toes is t. The allele for 5 toes is T.

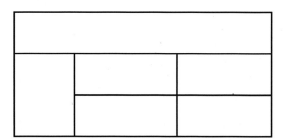

Fill in the probability table using the data from the Punnett square.

Genotype	Probability	Fraction	Percentage

Phenotype	Probability	Fraction	Percentage

How many babies should have 5 toes?

How many babies should have 3 toes?

When they are born, three babies have 3 toes and one has 5 toes. How do you explain this?

Qwitekutesnutes have 3 toes if their genotype is tt. What is this type of allele, t, called?

Qwitekutesnutes have 5 toes if their genotype is TT or Tt. What is this type of allele, T, called?

Pandia PRESS

Unit IV: Anatomy and Physiology
Exam Chapters 11–19

1. Multiple Choice

The organ(s) in plants where photosynthesis takes place:

○ Flowers

○ Leaves

○ Roots

○ Stems

○ All of the above

The process plants use to move food from the site where it is made is called

○ perspiration.

○ transpiration.

○ circulation.

○ translocation.

Your ears, eyes, mouth, and nose work most closely with which organ system?

○ respiratory

○ circulatory

○ nervous

○ brain

The birth process has three stages. The order they follow is:

○ labor, delivery of baby, delivery of placenta

○ delivery of placenta, labor, delivery of baby

○ labor, delivery of placenta, delivery of baby

○ delivery of baby, labor, delivery of placenta

Your skin and white blood cells fight infections together. How?

○ White blood cells form a protective layer on your skin.

○ Your skin funnels germs through your pores to your white blood cells, so they can destroy them.

○ Your skin prevents most germs from getting in, but when they do get in, the white blood cells destroy them.

○ When you get a cut, white blood cells rush to the site to form a clot over the cut.

When a plant breaks out of the seed and begins to grow, it is called

- ○ germination.
- ○ translocation.
- ○ fertilization.
- ○ sporophyte.

When a mosquito bites your hand, nerve cells in your hand send a signal telling your brain to smack it using

- ○ blood flow.
- ○ capillary action.
- ○ electrical signals.
- ○ hormones.

Where are blood cells made?

- ○ Inside your heart
- ○ Inside your bones
- ○ In the dermis layer of your skin
- ○ In your blood vessels

When you bend your knee, _____ connecting your bones and muscles push and pull them, while the _____ hold the bones together, and the _____ helps the bones glide against each other.

- ○ cartilage, ligaments, tendons
- ○ cartilage, tendons ligaments
- ○ ligaments, tendons, cartilage
- ○ tendons, ligaments, cartilage

Sperm are produced in male organs called

- ○ testes.
- ○ urethras.
- ○ ovaries.
- ○ cervix.

During ovulation

○ a baby is born.

○ a zygote is formed.

○ an egg is released into the fallopian tube.

○ the placenta is delivered.

The chemical messengers that send information back and forth between your organs are

○ blood.

○ electrical signals.

○ platelets.

○ hormones.

The part of your eye where images are formed is the

○ retina.

○ cornea.

○ iris.

○ pupil.

Blood is what type of tissue?

○ Ground tissue

○ Connective tissue

○ Vascular tissue

○ Epithelial tissue

Xylem and phloem are what type of tissue?

○ Ground tissue

○ Connective tissue

○ Vascular tissue

○ Epithelial tissue

Skeletal muscles are

○ involuntary muscles.

○ smooth muscles.

○ cardiac muscles.

○ voluntary muscles.

Bile is a chemical that

- ○ breaks down fat.
- ○ controls the sugar level in blood.
- ○ controls the production of white blood cells.
- ○ increases your heart rate.

_____ decreases the amount of glucose in your blood, and _____ increases the amount of glucose in your blood.

- ○ Thyroxin, adrenaline
- ○ Human growth hormone, estrogen
- ○ Insulin, glucagon
- ○ Oxytocin, thymosin

Once white blood cells have encountered a pathogen, the pathogen can't make you sick ever again. This is called being _____ to that pathogen.

- ○ immune
- ○ susceptible
- ○ homeostatic
- ○ hormonal

When you cut yourself, particles in your blood called platelets produce a chemical, which forms a net that traps blood cells and plasma, and forms a clot. This chemical is called

- ○ plasma.
- ○ oxytocin.
- ○ fibrin.
- ○ vaccine.

2. **Flower Parts and Fertilization.** Label the parts on the flower diagram below. Label the pistil, style, stigma, ovary, ovules, stamen, anther, and filament. Write, label, and draw the three steps of plant fertilization.

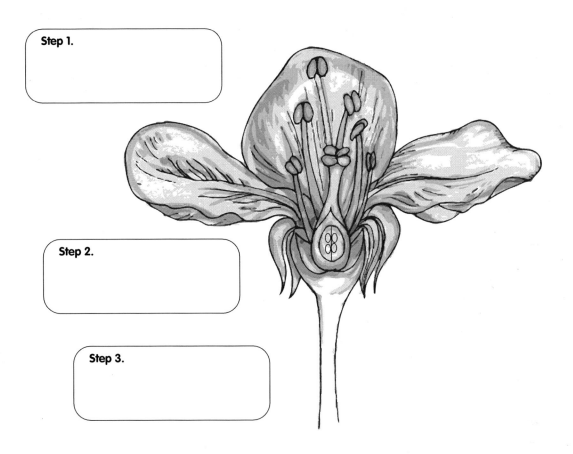

Step 1.

Step 2.

Step 3.

3. **Circulation.** Blood flows through arteries, veins, and capillaries. What is the difference between them? Fill in the blanks below.

_____ carry blood away from your heart to the rest of your body.

_____ are one cell thick, which allows the transport of small molecules across the walls of the membrane.

_____ carry blood to your heart from the rest of your body.

4. Homeostasis

What is homeostasis?

Describe how plants use the process of transpiration to maintain homeostasis.

Plants need to control their water balance because . . . (Name three reasons.)

1.)

2.)

3.)

Pandia PRESS

Your kidneys are important for your body maintaining homeostasis. What are your kidneys responsible for? How does that relate to transpiration in plants?

5. **Matching.** Write the letter in the space matching each organ system with the organs that make it up.

A. Nervous system   _____ skin, nails, hair

B. Integumentary system   _____ lymph nodes, lymphatic vessels, tonsils, thymus, spleen, Peyer's patches

C. Digestive system   _____ muscles

D. Urinary system   _____ brain, spinal cord, nerves

E. Reproductive system   _____ heart, arteries, veins, capillaries

F. Endocrine system   _____ salivary gland, esophagus, stomach, small intestine, liver, pancreas, gallbladder, large intestine, rectum, anus

G. Circulatory system   _____ bones, ligaments, cartilage, tendons

H. Respiratory system   _____ kidneys, ureters, bladder, urethra

I. Muscular system   _____ hypothalamus, endocrine glands, thymus, pancreas, testes OR ovaries

J. Skeletal system   _____ testes, scrotum, urethra, penis OR ovaries, fallopian tubes, uterus, cervix, vagina

K. Immune and lymphatic systems   _____ larynx, trachea, bronchi, lungs

6. **Organ systems.** Below is a list of purposes for each organ system. Write the name of the organ system from the list that best matches each purpose. Use each organ system only once. The immune and lymphatic systems are treated as one system.

Circulatory system   Endocrine system   Immune and lymphatic systems

Respiratory system   Reproductive system   Nervous system   Digestive system

Urinary system   Integumentary system   Muscular system   Skeletal system

_____ Works with the skeletal system so you can move

_____ Makes us

_____ Gives support and shape; works with the muscular system so you can move

_____ Protects the inside of the body, to send sensory information about touch to the brain

_____ Keep your body from getting sick

_____ Controls your other systems; stores memories and allows for thought

_____ Removes waste; balances the amount of fluid in your body

_____ Carries blood throughout your body

_____ Takes food and mashes it down into molecules that cells can transport across cell membranes

_____ Produces hormones

_____ Delivers oxygen from the air to your blood

7. **Extra Credit.** With a partner, use your body to show the approximate location of each of your organs. One point for each organ you correctly locate and name. There are a lot of organs, so you have the opportunity for a lot of extra credit.

Unit V: Evolution
Exam Chapters 20–23

1. Multiple Choice

When the first photosynthetic organisms evolved, they spewed a waste product,_____,into the air and seas. This waste product in the seas led to the _____that can now be seen in sedimentary rocks around the world. This waste product in the air led to the creation of the _____.

- ○ carbon dioxide, carbon-14, greenhouse effect
- ○ oxygen, banded iron formation, ozone layer
- ○ water, lightest layers, humidity
- ○ phosphorus-40, stratigraphy, radioactive isotopes

When biological evolution occurs, it happens to a(n)

- ○ organism.
- ○ individual.
- ○ population.
- ○ species.

The definition of population is

- ○ a group of organisms of the same species living in the same area.
- ○ a group of organisms that are genetically related so that they can reproduce, producing offspring that can also reproduce.
- ○ a group of living beings.
- ○ all the organisms in an area that interact with each other.

The definition of species is

- ○ a group of organisms living in the same area.
- ○ a group of organisms that are genetically related so that they can reproduce, producing offspring that can also reproduce.
- ○ a group of living beings.
- ○ all the organisms in an area that interact with each other.

A good index fossil

○ left a large number of fossils.

○ lived for a short period of geological time.

○ lived all over the world.

○ All of the above

○ 1 and 2 above

Genetic variation is caused by

○ mutation and genetic recombination.

○ beneficial and neutral adaptations.

○ asexual reproduction.

○ natural selection and reproductive isolation.

The case of the peppered moth, where dark moths were less likely to be eaten when tree bark darkened and therefore became more common than light-colored moths, is an example of

○ natural selection.

○ neutral adaptation.

○ beneficial adaptation.

○ All of the above

○ 1 and 3 above

The process that explains how all species of organisms came to be is called

○ natural selection.

○ genetic drift.

○ evolution.

○ sexual reproduction.

Some of the strongest evidence for evolution comes from the homologies between different species of organisms. The word *homologies* means what?

○ Differences

○ Similarities

○ Genetic variations

○ Relationships

The forearms of a dolphin and a lizard are similar. This is an example of

- ○ vestigial structures.
- ○ homologous structures.
- ○ chemical homology.
- ○ embryology.

Humans have a tailbone at the end of their spine but no tail. This bone is a _____ left over from a non-human ancestor in our distant past.

- ○ vestigial structure
- ○ homologous structure
- ○ chemical homology
- ○ genetic homology

The fossil record shows that

- ○ organisms have changed over time.
- ○ they started out simple and increased in complexity.
- ○ there are transitional fossils that show evolutionary links.
- ○ All of the above

With permineralization,

- ○ the organism is trapped in resin, which turns into the mineral amber.
- ○ a three-dimensional representation of the organism is fossilized as minerals fill in the space where the organism's body was.
- ○ an organism's soft tissue is replaced with minerals and creates a fossilized organism that is a rock.
- ○ an organism turn into a mineralized carbon trace.

Archaeopteryx fossils are transitional fossils that show the link between

- ○ birds and mammals.
- ○ birds and reptiles.
- ○ mammals and dinosaurs.
- ○ reptiles and amphibians.

The Principle of Superposition indicates that

- ○ the oldest fossils are at the top and the youngest fossils are at the bottom.
- ○ in a series of layered sedimentary rocks, the oldest layer is at the top and the youngest are at the bottom.
- ○ in a series of layered sedimentary rocks, the youngest layer is at the top and the oldest are at the bottom.
- ○ the youngest fossils are at the top and the oldest fossils are at the bottom.
- ○ 1 and 2
- ○ 3 and 4

You have grown up and become a world-famous biologist studying how saber-tooth tigers went extinct. While hiking in Alaska you find a fossil of a saber-tooth tiger that has become exposed by melting glaciers. What radioactive isotope would you use to find out how old it is?

- ○ C-14
- ○ P-40
- ○ U-235
- ○ O-18

2. True (T) or False (F). If false, fix the incorrect words to make it true.

_____ It is a fact that evolution occurs.

_____ It is a fact that birds evolved from dinosaurs.

_____ A scientific theory is based on the opinion of many scientists.

_____ A qwitekutesnute was born with a mutation such that a pocket of skin went from its underarms to its sides. This mutation enabled it to glide short distances between trees. This qwitekutesnute had lots of offspring, many with this trait. This means this trait is heritable.

_____ A harmful trait is selected for and will lead to more offspring with this trait.

_____ The half-life of a radioactive element is the amount of time it takes for half of the radioactive isotopes to turn into stable atoms.

_____ The more closely related two species are, the more similar their DNA is.

_____ The chemistry of all organisms is not homologous. They are made from the same six elements and the same types of molecules.

_____ Fossils are usually found in sedimentary rock.

_____ The Principle of Cross-Cutting Relations states that a geological feature is younger than anything it cuts across. This means that an igneous layer is older than the sedimentary layers because layers can only cut through a layer if it is already deposited.

3. Matching. Write each event in the textbox next to the correct date on the timeline.

multicellular life evolves	dinosaurs go extinct
mammals begin evolving into the many species of today	the first living organisms
plants colonize land	photosynthesis evolves
dinosaurs roam the earth	banded iron formation
animals colonize land	eukaryotic cells evolve

TIMELINE

4.5 –3.8 billion years ago _____

3 billion years ago _____

2.5–2 billion years ago _____

1.85 billion years ago _____

1.5 billion years ago _____

450 million years ago _____

420 million years ago _____

230 million years ago _____

65 million years ago _____

after 65 million years ago _____

4. Vocabulary Match. Match each vocabulary term with the best definition.

A. Natural selection

_____ an inherited variation that affects the survival of organisms with that trait

B. Genetic drift

_____ when a new species evolves from existing species

C. Speciation

_____ the random change in the frequency of alleles in a population due to chance events in small populations

D. Overproduction

_____ occurs when organisms have a better or a worse chance of survival because of their traits

E. Genetic variation

_____ where reproduction is stopped between a population of organisms and other populations within the species

F. Adaptation

_____ genetic changes to populations

G. Genetic recombination

_____ more organisms are born than the environment can support

H. Reproductive isolation

_____ variation of traits between members in a population caused by allelic differences

I. Evolution

_____ when homologous chromosomes crossover one another and exchange pieces of DNA

5. Extra Credit. A qwitekutesnute is born with opposable thumbs (thumbs like you have but your dog and cat do not). Choose whether you think this trait is beneficial or harmful. Write a paragraph supporting your choice. Beneficial trait or harmful trait could be correct. There are no points for your choice. The points are for the support of your choice.

Unit VI: Ecology
Exam Chapters 24–27

1. Multiple Choice

The most important abiotic factor affecting where different terrestrial biomes develop is

- ○ the amount of sunlight.
- ○ climate.
- ○ biodiversity.
- ○ longitude.

Climate change is a problem for organisms because

- ○ the climate on earth has always been the same and organisms can't change if it changes.
- ○ the size of the oceans will decrease, there will be more land, and there will be nowhere for all the organisms living in the sea to go.
- ○ the climate is changing so fast that organisms do not have time to adapt at the rate of change.
- ○ the biomes will be affected the same, so there is nowhere to go.

Tapeworms live in the digestive tract of other animals. When the host animal eats, the tapeworm eats too. The relationship between the tapeworm and the animal it lives in is called _____ and it is a type of _____.

- ○ parasitism, symbiosis
- ○ mutualism, symbiosis
- ○ commensalism, parasitism
- ○ predation, habitat

The greenhouse effect is most affected by which cycle?

- ○ The water cycle
- ○ The nitrogen cycle
- ○ The phosphorus cycle
- ○ The carbon cycle

When deer eat grass, there is _____ energy going from grass to the deer.

- ○ more
- ○ less
- ○ the same amount
- ○ predatory

When a local population of a species goes extinct, it is called

- ○ poaching.
- ○ desertification.
- ○ loss of habitat.
- ○ extirpation.

The uneven heating of the earth leading to differences in temperature based on latitude is caused by

- ○ longitude.
- ○ climate.
- ○ the earth's tilt.
- ○ precipitation.

In a community, different species of animals have evolved different strategies for using the same limited resources; this decreases competition. Scientists say that each population has its own _____ .

- ○ commensalism
- ○ abiotic interactions
- ○ niche
- ○ symbiosis

Pollution often occurs when the chemistry of an environment is changed. Common pollutants are

- ○ pesticides.
- ○ herbicides.
- ○ fertilizers.
- ○ acids.
- ○ All of the above
- ○ None of the above

The abiotic factor that has the biggest effect on who lives where in the aquatic zone is

- ○ longitude.
- ○ latitude.
- ○ light.

Global warming is the increase of the average world temperature because of

- ○ an increase of heat-trapping molecules such as water vapor in the atmosphere.
- ○ an increase of heat-trapping molecules such as oxygen in the atmosphere.
- ○ an increase of heat-trapping molecules such as carbon dioxide in the atmosphere.
- ○ an increase of heat-trapping molecules such as phosphates in the atmosphere.

Overfishing is when so many fish are caught there are not enough to sustain the population. Overfishing is a problem for over _____ of all species of fish.

- ○ 50%
- ○ 100%
- ○ 25%
- ○ 10%

The abiotic components in the environment are

- ○ those that are the non-living chemical and physical parts.
- ○ those that are living or were once living.
- ○ all the organisms that live in an area.
- ○ the parts that cause pollution.

Activities that change an area from one where wildlife lives to one where it does not is called

- ○ desertification.
- ○ extirpation.
- ○ extinction.
- ○ loss of habitat.

Competition is the weakest between organisms with _____ niches.

- ○ the same
- ○ similar
- ○ the most different
- ○ symbiotic

A tiger's stripes are a type of camouflage called

○ disruptive coloring.

○ aposematic coloring.

○ mimicry.

○ bioluminescence.

The most intense competition is between

○ predators and their prey.

○ the same species of organisms.

○ carnivores.

○ species with very similar niches.

Human causes for the increase of heat-trapping molecules in the atmosphere are

○ exhaust fumes from cars.

○ factories that burn oil and coal.

○ electricity use in homes.

○ deforestation.

○ All of the above

A carbon footprint means the things that _____ carbon dioxide. It helps the earth when people _____ their carbon footprint.

○ reduce, increase

○ photosynthesize, eat

○ recycle, reuse

○ increase, reduce

When fertilizers wash into aquatic ecosystems, they make plants grow better. The increase in plant growth uses up oxygen that is dissolved in the water. This decrease in oxygen can kill other organisms. When this happens, it is called

○ extirpation.

○ eutrophication.

○ carbon footprint.

○ desertification.

2. Food web. Draw a food web on the watery scene below. Add to the picture at least nine organisms that could consume and/or be consumed in this biome. Some examples you can draw include fish, insect, flower, bird of prey, songbird, waterfowl, snail, person, snake, seeds, tree nuts, frog, rat, alligator, etc. Draw arrows going away from an organism to anything it might be eaten by. Write a P above organisms that are producers.

3. Biomes. Pictured on the opposite page are seven types of biomes you studied in this unit. Below, there are three text boxes with a list of biome characteristics. First, title each biome pictured. Then, take one characteristic from each text box and match it with the biome best described by that characteristic. Write that characteristic next to the biome picture. (One is done for you.) Each biome gets one characteristic from each text box.

---

1. Water everywhere

2. Between 23.5° and 50° latitude in both hemispheres

3. A wet and a dry season; during the dry season fires are common

4. Very low amount of precipitation

5. 50° N and 60° N latitude

6. Cold temperatures with permafrost

7. At or near the equator between 23.5° N and 23.5° S latitude

---

1. Gets the most rain; has the most trees

2. Grasses with scattered trees

3. Plant roots are good at absorbing water; leaves are good at preventing water loss

4. ~~Plants and animals are affected by the amount of sunlight received~~

5. Treeless with short plants

6. Evergreen trees shaped so snow falls off them without breaking branches

7. Trees are deciduous

---

1. Small, burrowing animals with specialized kidneys

2. Animals have adaptations for living in trees, such as prehensile tails

3. Migrating or hibernating animals with shorter extremities

4. Some animals live in a subnivian zone in winter; others hibernate (this is common in two biomes)

5. Grazing herbivores

6. Animals have sleek, bullet-shaped bodies or can attach themselves to surfaces

---

___Aqueous___ Biome

1.

2. Plants and animals are affected by the amount of sunlight received

3.

_____ Biome

1.

2.

3.

_____ Biome

1.

2.

3.

_____ Biome

1.

2.

3.

_____ Biome

1.

2.

3.

_____ Biome

1.

2.

3.

_____ Biome

1.

2.

3.

4. Fill in the blank. Complete the essay using the words from the text box to the right of each paragraph.

The water cycle explains how water cycles through the environment. If we start with a drop of water in a lake, that drop _____ when the sun warms the water up. This water molecule goes up into the sky, where it _____ and forms a part of a cloud. It doesn't stay there forever, though; this water molecule _____ back to the earth in a raindrop. Once it hits the ground, it _____ into the ground. Tree roots in the ground absorb the drop using _____. This drop doesn't stay in the tree forever; soon enough it evaporates from the leaves by _____ and it is back up to the clouds for the little drop.

osmosis
condenses
percolates
evaporates
transpiration
precipitates

During the water cycle, a water molecule is always in the same molecular form, with two hydrogen atoms bonded to one oxygen atom. During the nitrogen cycle, nitrogen changes the molecules it is in, but it still remains nitrogen. Nitrogen gets into the air by _____ in the soil that release nitrogen into the air. The nitrogen floats around until _____ change it into nitrogen that organisms can use. In the soil there are also consumers called _____; they break down molecules in dead organisms and release nitrogen into the soil. The nitrogen _____ into plants or is _____ by animals, which make _____ and _____ with it.

diffuses
DNA
proteins
denitrifying bacteria
decomposers
eaten
nitrogen-fixing bacteria

Phosphorus is also needed to make _____. Phosphorus comes from the _____ of rocks. Some organisms get their phosphorus from what is dissolved in the water; others get it by _____ it or taking it up through roots with _____.

DNA
diffusion
weathering
eating

The molecules that make organisms have a lot of carbon in them. In addition, it is the carbon cycle that feeds life on Earth. Plants make their own food with _____. Animals get this food by _____ plants. Both plants and animals get the energy they need with _____.

respiration
photosynthesis
eating

5. Extra Credit: What biome do you live in? Describe to someone the climate of your biome, the native organisms found in your biome and the seasons and environmental factors affecting them, and any environmental problems your biome is facing.

Unit VII: Classification

Unit Exam Chapters 28–32

1. Multiple Choice. Each of the following multiple choice questions gives one or more traits. Choose the organism that is described by the trait(s).

I am a prokaryote who is classified based on differences in my RNA. What am I?

- ○ Bacteria
- ○ Archaea
- ○ Protista
- ○ Fungi
- ○ Plantae
- ○ Animalia

I am a unicellular prokaryote who is classified by shape, feeding strategy, and the conditions in which I grow. What am I?

- ○ Bacteria
- ○ Archaea
- ○ Protista
- ○ Fungi
- ○ Plantae
- ○ Animalia

I am a multicellular, eukaryotic autotroph. What am I?

- ○ Bacteria
- ○ Archaea
- ○ Protista
- ○ Fungi
- ○ Plantae
- ○ Animalia

I am a multicellular, eukaryotic heterotroph. I do not have cell walls. What am I?

- ○ Bacteria
- ○ Archaea
- ○ Protista
- ○ Fungi
- ○ Plantae
- ○ Animalia

I am a eukaryotic chemotroph. I am immobile. What am I?

- ○ Bacteria
- ○ Archaea
- ○ Protista
- ○ Fungi
- ○ Plantae
- ○ Animalia

I am a eukaryote that lives in watery places. I am usually unicellular. What am I?

- ○ Bacteria
- ○ Archaea
- ○ Protista
- ○ Fungi
- ○ Plantae
- ○ Animalia

I am in kingdom Plantae. I am small with no vascular tissue. I reproduce using spores. What am I?

- ○ Angiosperm
- ○ Bryophyte
- ○ Gymnosperm
- ○ Pterophyte

I am in kingdom Plantae. I have vascular tissue and reproduce with seeds on cones. What am I?

- ○ Angiosperm
- ○ Bryophyte
- ○ Gymnosperm
- ○ Pterophyte

I am in kingdom Plantae. I have vascular tissue and reproduce with seeds contained in flowers. What am I?

- ○ Angiosperm
- ○ Bryophyte
- ○ Gymnosperm
- ○ Pterophyte

I am in kingdom Plantae. I have vascular tissue and reproduce with spores. What am I?

- ○ Angiosperm
- ○ Bryophyte
- ○ Gymnosperm
- ○ Pterophyte

I am in kingdom Animalia. I have a backbone. What am I?

- ○ Mollusca
- ○ Arthropoda
- ○ Porifera
- ○ Platyhelminthes
- ○ Annelida
- ○ Nematoda
- ○ Cnidaria
- ○ Chordata
- ○ Echinodermata

I am in kingdom Animalia. I have a shell and a muscular foot I use to move. What am I?

- ○ Mollusca
- ○ Arthropoda
- ○ Porifera
- ○ Platyhelminthes
- ○ Annelida
- ○ Nematoda
- ○ Cnidaria
- ○ Chordata
- ○ Echinodermata

I am in kingdom Animalia. I have jointed legs and an exoskeleton. What am I?

- ○ Mollusca
- ○ Arthropoda
- ○ Porifera
- ○ Platyhelminthes
- ○ Annelida
- ○ Nematoda
- ○ Cnidaria
- ○ Chordata
- ○ Echinodermata

I am in kingdom Animalia. I am a parasitic worm. I have a long, threadlike body. What am I?

- ○ Mollusca
- ○ Arthropoda
- ○ Porifera
- ○ Platyhelminthes
- ○ Annelida
- ○ Nematoda
- ○ Cnidaria
- ○ Chordata
- ○ Echinodermata

I am in kingdom Animalia. I am a sponge. I have pores all over my body. What am I?

- ○ Mollusca
- ○ Arthropoda
- ○ Porifera
- ○ Platyhelminthes
- ○ Annelida
- ○ Nematoda
- ○ Cnidaria
- ○ Chordata
- ○ Echinodermata

I am in kingdom Animalia. I am an earthworm. I have a segmented body with male and female parts on it. What am I?

- ○ Mollusca
- ○ Arthropoda
- ○ Porifera
- ○ Platyhelminthes
- ○ Annelida
- ○ Nematoda
- ○ Cnidaria
- ○ Chordata
- ○ Echinodermata

I am in kingdom Animalia. I live in the marine biome. I have tough, spiny skin and a radial body with five sections. What am I?

- ○ Mollusca
- ○ Arthropoda
- ○ Porifera
- ○ Platyhelminthes
- ○ Annelida
- ○ Nematoda
- ○ Cnidaria
- ○ Chordata
- ○ Echinodermata

I am in kingdom Animalia. I am a jellyfish. I have tentacles with stinging cells around my mouth that I use to paralyze my prey. What am I?

- ○ Mollusca
- ○ Arthropoda
- ○ Porifera
- ○ Platyhelminthes
- ○ Annelida
- ○ Nematoda
- ○ Cnidaria
- ○ Chordata
- ○ Echinodermata

I am in kingdom Animalia. I am a parasitic worm with a flat body. What am I?

- ○ Mollusca
- ○ Arthropoda
- ○ Porifera
- ○ Platyhelminthes
- ○ Annelida
- ○ Nematoda
- ○ Cnidaria
- ○ Chordata
- ○ Echinodermata

The three types of mammals are

- ○ placental, marsupial, chordata.
- ○ myriapoda, hexapoda, monotreme.
- ○ monotreme, placental, marsupial.
- ○ myriapoda, placental, marsupial.

2. **Matching.** Phylum Chordata has five classes. Match the class with the description.

Fish ○	○	Ectotherm, lays eggs, dry scaly skin
Amphibian ○	○	Endotherm, mammary glands, fur
Reptile ○	○	Lives in water, breathes through gills
Bird ○	○	Ectotherm, goes through metamorphosis, moist skin
Mammal ○	○	Endotherm, lays eggs, has feathers

3. **Short Answers**

What evidence is used to classify organisms?

What are the three domains called?

4. **Vocabulary match.** Match the word with the definition that best fits.

Taxonomy ○   ○ Regulate their body temperature internally

Hyphae ○   ○ The evolutionary history of species

Ectotherm ○   ○ Formed by a symbiotic relation between algae and a
                                                      fungus

Cladistics ○   ○ Organism without a backbone

Phylogeny ○   ○ The branch of biology that classifies and names organisms

Endotherm ○   ○ Organism with a backbone

Dichotomous key ○   ○ Classification system based on phylogeny

Lichen ○   ○ Method of identifying organisms based on a series of
                                                      questions with two choices

Vertebrate ○   ○ Regulate body temperature by exchanging heat with
                                                      their environment

Invertebrate ○   ○ Structure used by fungi to get food

5. **Classification.** On the right side is the classification for a vampire squid. Fill in the left side of the table with the names of the eight levels of classification.

	Eukarya
	Animalia
	Mollusca
	Cephalopoda
	Vampyromorphida
	Vampyroteuthidae
	Vampyroteuthis
	infernalis

What is the scientific name of the vampire squid? What is this system of naming organisms called?

6. Cladogram. Fill in the cladogram below. Write the shared traits on the horizontal lines and the organisms at the top of the vertical lines.

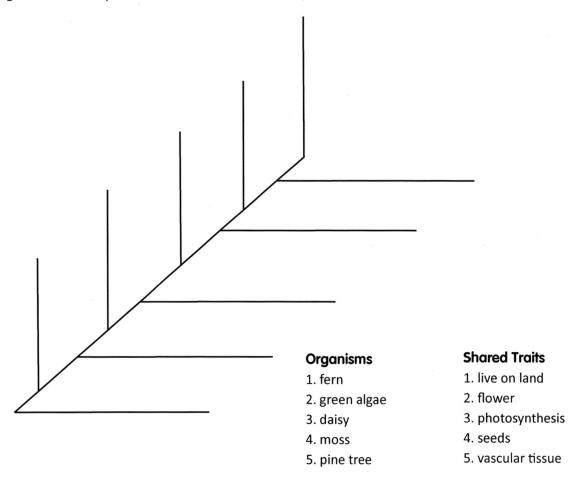

Organisms
1. fern
2. green algae
3. daisy
4. moss
5. pine tree

Shared Traits
1. live on land
2. flower
3. photosynthesis
4. seeds
5. vascular tissue

7. Extra Credit

There are four main classes of arthropods. What are they called?

Name 1 organism from each class.

What is 1 characteristic specific to each class?

Appendix B

How to Use Punnett Squares

Two qwitekutesnutes are going to have babies. Some qwitekutesnutes have three toes and some have five toes. The allele for five toes, T, is dominant over the allele for three toes, t. The mother has the genotype Tt, and the father has the genotype tt. When you look at the possible allele combinations for two parents, it is called a *cross*. The cross for these two parents is Tt x tt. This is the genotype for each parent. The phenotype for this trait is five toes for the mother who is Tt and three toes for the father who is tt.

A Punnett square shows all the possible allele combinations for this trait that parents can pass on to their offspring based on the alleles the parents have. Choose one of the parents (it doesn't matter which one) and write their genotype along the top of the Punnett square. Along the other side do the same for the other parent. Fill in each box showing the combination of those two alleles.

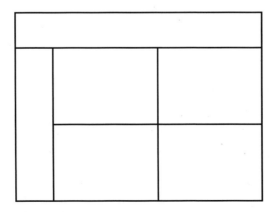

Your Punnett square will look something like this:

	T	t
t	Tt	tt
t	Tt	tt

The information can be transferred to a probability table. A probability table tells the likely genotype and phenotype distribution for the alleles, based on the genotype of the parents. As you can see from the Punnet square, there are only two possible genotypes that can occur in the babies of these parents: Tt and tt. This also means both phenotypes are possible: five toes and three toes. Complete the probability table for these genotypes and phenotypes:

1	2	3	4	5	6	7	8
Genotype	Probability	Fraction	Percentage	Phenotype	Probability	Fraction	Percentage
Tt				5 toes			
tt				3 toes			

Completing a Probability Table

1. Genotype column: List the possible genotypes from the Punnett square; the order does not matter. (This has already been done for you.)

2. Probability column: Count the number of squares in which that genotype appears in the Punnett square. Relate it to the total number of squares in the Punnett square.

3. Fraction column: Rewrite the genotype probability as a fraction.

4. Percentage: Use a calculator, or the one in your head, to divide the fraction, ex. 2/4 → 2 ÷ 4 = 0.50. Multiply the answer by 100 to get the genotype percentage, ex. 0.50 × 100 = 50%

5. Phenotype: T is the dominant allele and t is the recessive allele. The genotype Tt results in the phenotype 5 toes. The genotype tt results in the phenotype 3 toes. Write a very brief description of the phenotype in each box. (This has already been done for you.)

6. Probability column: Count the number of squares that phenotype appears in the Punnett square. Relate it to the total number of squares in the Punnett square.

7. Fraction column: Rewrite the phenotype probability as a fraction.

8. Percentage: Calculate and report the phenotype percentage.

Tt   tt   tt

Your probability table should look like this:

Genotype	Probability	Fraction	Percentage	Phenotype	Probability	Fraction	Percentage
Tt	2 in 4	2/4	50%	5 toes	2 in 4	2/4	50%
tt	2 in 4	2/4	50%	3 toes	2 in 4	2/4	50%

What this type of analysis does not tell you: Punnett squares and probability tables are tools used to predict the most likely outcomes based on the data. As we saw in the chapter on meiosis, there are many possible outcomes for offspring during fertilization. Organisms get their alleles from their parents. The exact combination of alleles an organism gets from its parents is random and cannot be foretold.

Appendix C

Essay Worksheet

Topic

Topic Sentence

Main Idea #1	Main Idea #2	Main Idea #3

Introduction Paragraph (from topic sentence and main ideas)

Body Paragraph 1 (from main idea #1)

Body Paragraph 2 (from main idea #2)

Body Paragraph 3 (from main idea #3)

Conclusion Paragraph (from topic sentence and main ideas)

Pandia PRESS

Research

Appendix D

Attribution of Sources

When conducting research for writing assignments, you will be reading books and passages written by other people. If you want to use the writings of others in your research report, you will need to *paraphrase* the work or *quote* the author and then attribute (give credit for) the work to the author. Proper attribution of sources is very important and helps you avoid plagiarism. *Plagiarism* is presenting someone else's work as your own or not properly attributing an idea to the author. Plagiarism can be a serious offense. At many colleges and high schools, students receive a failing grade if they plagiarize. Also, you should be aware that professors have sophisticated software and resources to assist them in detecting plagiarism.

PARAPHRASING AND QUOTING

Paraphrasing is restating a passage and conveying its meaning with different words. To paraphrase correctly, you need to restate the original author's ideas in your own words. Simply changing a few words in a sentence is not paraphrasing. The best way to paraphrase is to begin by thoroughly reading the passage you want to paraphrase. Then close the book and rewrite the idea without looking at the original work. Be sure to cite all of the authors and their works from which you borrowed ideas in the bibliography at the end of your report (see bibliography examples on the next page).

A *quote* is the exact words of the author placed in quotation marks. When using a quote, state the words *exactly* as the author did. Most of the time it is more appropriate to paraphrase an author than to directly quote him or her. But occasionally you will want to use a quote. You might want to use a quote when the words of the author are particularly powerful, when you are quoting a line in literature, or when using the words of a famous person. For example:

When Rousseau said, "Man is born free, and everywhere he is in chains," he implied that people are hindered by the limitations of their government.

BIBLIOGRAPHY

A bibliography is a list of the books, articles, Internet sites, and audiovisuals from which you gathered information when preparing your report. When do you need to cite a source in a bibliography? Basically, you need to cite any source from which you borrow an idea, use direct quotes, or write a paraphrase in your report. You do not need to cite a source when the knowledge is common knowledge or a well-known fact.

When writing a bibliography, you should . . .

• Put the sources in alphabetical order by the author's last name or by the first word of the title if there is no author (not counting "a," "an," or "the").

• Indent the second line of an entry if you need to use more than one line.

• Skip a line after each entry.

• Underline the title of a book or magazine (or use italics if typing).

- List the authors in the order they are listed on the title page when there is more than one author.

- List the title of an article from a newspaper of encyclopedia before the name of the newspaper or encyclopedia. Put titles of articles in quotation marks.

BIBLIOGRAPHY EXAMPLES*

BOOK:
Author's last name, first name. *Title of book*. Place of publication: Publisher, copyright year.

Example:
Carson, Rachel. *Silent Spring*. Boston: Mariner Books, 1962.

ENCYCLOPEDIA ARTICLE WITHOUT AN AUTHOR:
"Title of article." *Name of encyclopedia*. Edition number. Copyright year.

Example:
"Endangered Species." *World Book Encyclopedia*. 10th ed. 1999.

MAGAZINE OR NEWSPAPER ARTICLE:
Article author's last name, first name. "Title or headline of article." *Name of magazine or newspaper.* Date of magazine or newspaper, section and page.

Example:
Foster, Joanna. "Parasites Can Be Good for You (Seriously)." *New York Times.* May 22, 2012, C1

INTERNET ADDRESS:
Author's last name, first name. "Title of item." Date of document or download. http://address
If there is no author cited, then begin with the title.

Example:
"Species." June 4, 2005. academickids.com/encyclopedia/index.php?title=Species

FILM:
Title of film. Director. Distributor, year of release.

Example:
The 11th Hour. Dir. Leila Conners. Appian Way, 2007.

* According to the MLA. Gibaldi, Joseph. *MLA Handbook for Writers of Research Papers*. 6th ed. New York: The Modern Language Association of America, 2003.